Effortless Algebra

maranGraphics™

&

THOMSON

™

COURSE TECHNOLOGY

Professional ■ Technical ■ Reference

MARAN ILLUSTRATED™ Effortless Algebra

Distributed in the U.S. and Canada by Thomson Course Technology PTR. For enquiries about Maran Illustrated™ books outside the U.S. and Canada, please contact maranGraphics at international@maran.com

For U.S. orders and customer service, please contact Thomson Course Technology at 1-800-354-9706. For Canadian orders, please contact Thomson Course Technology at 1-800-268-2222 or 416-752-9448.

ISBN: 1-59200-942-5

Library of Congress Catalog Card Number: 2005926159

Printed in the United States of America

05 06 07 08 09 BU 10 9 8 7 6 5 4 3 2 1

Trademarks

Important

Copies

Educational facilities, companies, and organizations located in the U.S. and Canada that are interested in multiple copies of this book should contact Thomson Course Technology PTR for quantity discount information. Training manuals, CD-ROMs, and portions of this book are also available individually or can be tailored for specific needs.

maranGraphics™

maranGraphics Inc.
5755 Coopers Avenue
Mississauga, Ontario
L4Z 1R9
www.maran.com

THOMSON

COURSE TECHNOLOGY
Professional ■ Technical ■ Reference

Thomson Course Technology PTR, a division of Thomson Course Technology
25 Thomson Place ■ Boston, MA 02210 ■ http://www.courseptr.com

maranGraphics is a family-run business

At **maranGraphics**, we believe in producing great consumer books–one book at a time.

Each maranGraphics book uses the award-winning communication process that we have been developing over the last 30 years. Using this process, we organize photographs and text in a way that makes it easy for you to learn new concepts and tasks.

We spend hours deciding the best way to perform each task, so you don't have to! Our clear, easy-to-follow photographs and instructions walk you through each task from beginning to end.

We want to thank you for purchasing what we feel are the best books money can buy. We hope you enjoy using this book as much as we enjoyed creating it!

Sincerely,

The Maran Family

We would love to hear from you! Send your comments and feedback about our books to family@maran.com

To sign up for sneak peeks and news about our upcoming books, send an e-mail to newbooks@maran.com

Please visit us on the Web at:

www.maran.com

CREDITS

Author:
maranGraphics Development Group

Content Architect:
Ruth Maran

Technical Consultant:
Alistair Savage Ph.D.

Project Manager:
Judy Maran

Copy Development Director:
Kelleigh Johnson

Copy Developer:
Andrew Wheeler

Layout Designers:
Richard Hung
Sarah Kim

**Front Cover Image and
Overview Designer:**
Russ Marini

Indexer:
Andrew Wheeler

Post Production:
Robert Maran

**President,
Thomson Course Technology:**
David R. West

**Senior Vice President of
Business Development,
Thomson Course Technology:**
Andy Shafran

**Publisher and General Manager,
Thomson Course Technology PTR:**
Stacy L. Hiquet

**Associate Director
of Marketing,
Thomson Course Technology PTR:**
Sarah O'Donnell

**National Sales Manager,
Thomson Course Technology PTR:**
Amy Merrill

**Manager of Editorial Services,
Thomson Course Technology PTR:**
Heather Talbot

ACKNOWLEDGMENTS

Thanks to the dedicated staff of maranGraphics, including
Richard Hung, Kelleigh Johnson, Jill Maran Dutfield, Judy Maran,
Robert Maran, Ruth Maran, Russ Marini and Andrew Wheeler.

Finally, to Richard Maran who originated the easy-to-use graphic format
of this guide. Thank you for your inspiration and guidance.

Alistair Savage, Ph.D.

Alistair Savage is a Postdoctoral Fellow at the Fields Institute for Research in Mathematical Sciences and the Department of Mathematics at the University of Toronto. He completed his undergraduate degree in honours mathematics and physics at the University of British Columbia and his Ph.D. in mathematics at Yale University. After obtaining his doctorate, Alistair conducted mathematical research at the Max-Planck-Institut für Mathematik in Bonn, Germany before starting his current position in Toronto.

Alistair is an active researcher in the field of algebra and his work has been published in leading mathematical journals.

He has lectured extensively in Canada, the United States, Germany and Japan and has taught courses in many areas of mathematics including algebra and calculus at both Yale University and the University of Toronto.

Table of Contents

Table of Contents

Table of Contents

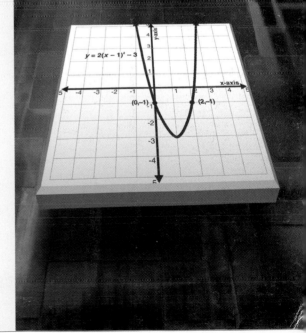

Chapter 1

Welcome to algebra boot camp! Before you jump into algebra, you need to understand the basics. There are new terms, iron-clad rules and a few properties with which you should become familiar. Chapter 1 provides you with a solid grounding in the essentials of algebra.

Algebra Basics

In this Chapter...

Common Algebra Terms

Like any other specialized pursuit, algebra has a language all its own, but don't be intimidated. Not long ago, terms like Web site and e-mail would have left most people scratching their heads in confusion. Today, however, these terms have entered our vocabulary and we use them without a second thought.

In the study of algebra, you will hear terms like coefficient, integer and polynomial thrown around quite often. These terms are just simple ways of communicating more complicated ideas. Once you understand these terms and see them in context, you should not have any trouble remembering them.

Absolute Value

The absolute value of a number is always the positive value of the number, whether the number is positive or negative. For example, the absolute value of both 5 and -5 is 5. An absolute value is indicated by two thin vertical lines (| |) around a number, such as $|5|$.

Coefficient

The coefficient is a number in front of a variable. For example, 5 is the coefficient in the expression $5x$.

Expression

An expression is a mathematical statement that does not contain an equals sign ($=$). An expression can contain numbers, variables and/or operators, such as $+$, $-$, $\times$ and $\div$. For example, $x + y$ and $5x - 2y$ are expressions.

Equation

An equation is a mathematical statement containing an equals sign ($=$). The equals sign indicates that the two sides of the equation are equal. An equation whose highest exponent in the equation is one, such as $x + y = 10$, is known as a linear equation. An equation whose highest exponent in the equation is two, such as $x^2 + 4x = 16$, is known as a quadratic equation.

Formula

A formula is a statement expressing a general mathematical truth and can be used to solve or reorganize mathematical problems. For example, $(a + b)^2$ always equals $(a + b)(a - b)$, so $(a + b)^2 = (a + b)(a - b)$ is a formula you can use to work with problems.

Exponent

An exponent, or power, appears as a small number above and to the right of a number, such as 3 in 2^3. An exponent indicates the number of times a number is multiplied by itself. For example, 2^3 equals $2 \times 2 \times 2$. The number to which the exponent is attached, such as 2 in 2^3, is called the base.

Fraction

A fraction, such as $\frac{1}{2}$ or $\frac{3}{4}$, is a division problem written with a fraction bar ($-$) instead of with a division sign ($\div$). For example, you can write $3 \div 4$ as $\frac{3}{4}$. A fraction has two parts—the top number in a fraction is called the numerator and the bottom number in a fraction is called the denominator.

Factor

Factors are numbers or terms that are multiplied together to arrive at a specific number. For example, 3 and 4 are factors of 12. When you factor an expression in algebra, you break the expression into pieces, called factors, that you can multiply together to give you the original expression.

Inequality

An inequality is a mathematical statement in which one side is less than, greater than, or possibly equal to the other side. Inequalities use four different symbols—less than ($<$), less than or equal to ($\leq$), greater than ($>$) and greater than or equal to ($\geq$). For example, the inequality $10 < 20$ states that 10 is less than 20.

Prime Number

A prime number is a positive number that you can evenly divide only by itself and the number 1. For example, $2, 3, 5, 7, 11$ and 13 are examples of prime numbers. The number 1 is not considered a prime number.

Radical

A radical is a symbol ($\sqrt{}$) that tells you to find the root of a number. In algebra, you will commonly find the square root of numbers. To find the square root of a number, you need to determine which number multiplied by itself equals the number under the radical sign. For example, $\sqrt{25}$ equals 5.

Integer

An integer is a whole number or a whole number with a negative sign ($-$) in front of the number. For example, $-3, -2, -1, 0, 1, 2$ and 3 are examples of integers.

Matrix

$$\begin{bmatrix} 1 & 7 & 5 \\ 2 & 6 & 4 \\ 8 & 3 & 9 \end{bmatrix}$$

A matrix is a collection of numbers, called elements, which are arranged in horizontal rows and vertical columns. The collection of numbers is surrounded by brackets or parentheses. Matrices is the term used to indicate more than one matrix.

Terms

Terms are numbers, variables or a combination of numbers and variables, which are separated by addition ($+$), subtraction ($-$), multiplication ($\times$) or ($\cdot$), or division ($\div$) signs. For example, the expression xy contains one term, whereas the expression $x + y$ contains two terms.

Solve

When you are asked to solve a problem, you need to find the answer or answers to the problem. For example, when you solve the equation $2x - 6 = 0$, you determine that x equals 3.

System of Equations

A system of equations is a group of two or more related equations. You are often asked to find the value of each variable that solves all the equations in the group.

Variable

A variable is a letter, such as x or y, which represents an unknown number. For example, if x represents Emily's age, then $x + 5$ represents the age of Emily's sister who is 5 years older.

Polynomial

A polynomial is an expression that consists of one or more terms, which can be a combination of numbers and/or variables, that are added together or subtracted from one another. For example, $2x^2 + 3x$ is a polynomial. A polynomial with only one term, such as $5x$, is called a monomial. A polynomial with two terms, such as $5x + 7$, is called a binomial. A polynomial with three terms, such as $2x^2 + 5x + 7$, is called a trinomial.

Classifying Numbers

There are an awful lot of numbers—ranging from familiar numbers used to count everyday things, like two apples, to huge abstract amounts that one can barely comprehend, such as Bill Gates' net worth. That vast range, however, only encompasses a miniscule fraction of the numbers that exist. Ironically, you cannot count the number of numbers. That being the case, mathematicians have classified numbers into more manageable groups, or number systems.

You will already be familiar with some of these groups, like positive, negative and odd numbers, while others may be new concepts, such as prime numbers, integers and irrational numbers. It is good to be on friendly terms with all of these number systems as they are all utilized in algebra. Many have quirky traits and relationships to other numbers that will come in very handy as you learn more about them.

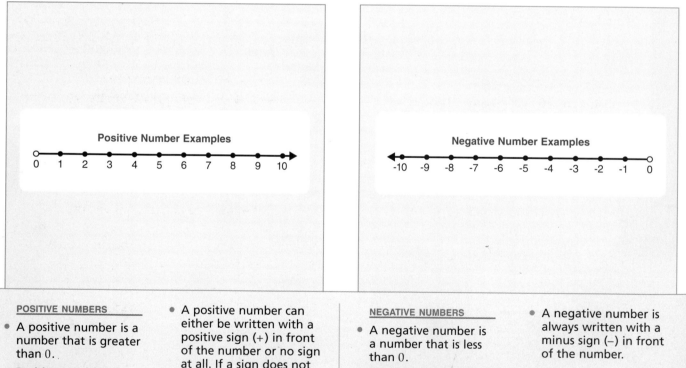

Positive Number Examples

Negative Number Examples

POSITIVE NUMBERS

- A positive number is a number that is greater than 0.
- Positive numbers become larger the farther they are from zero. For example, 64 is larger than 22.

- A positive number can either be written with a positive sign (+) in front of the number or no sign at all. If a sign does not appear in front of a number, consider the number to be a positive number. For example, +4 can also be written as 4.

NEGATIVE NUMBERS

- A negative number is a number that is less than 0.
- Negative numbers become smaller the farther they are from zero. For example, −64 is smaller than −22.

- A negative number is always written with a minus sign (−) in front of the number.

Tip

How is 0 classified?

Talk about a loaded question. Zero is a number with a lot of peculiar properties that mathematicians have been debating for centuries. For our purposes, 0 can be classified as an even number because it can be evenly divided by 2. In terms of being classified as positive or negative, 0 sits in a numeric limbo between the groups and is considered neither positive nor negative.

Practice

$a+b=c$

Define each of the following statements as true or false. You can check your answers on page 250.

1) 5 is a positive, prime and composite number.

2) 3 is a positive, odd and prime number.

3) –20 is a negative, odd and composite number.

4) –7 is a negative, odd and prime number.

5) 24 is a positive, even and composite number.

6) 43 is a positive, odd and composite number.

Even Number Examples

2, 4, 6, 8, 10, –2, –4, –6, –8, –10

Odd Number Examples

1, 3, 5, 7, 9, –1, –3, –5, –7, –9

Prime Number Examples

2, 3, 5, 7, 11, 13, 17, 19, 23, 29

Composite Number Examples

4, 6, 8, 10, 12, 14, 15, 16, 18

EVEN NUMBERS

- An even number is a number that you can evenly divide by 2.

ODD NUMBERS

- An odd number is a number that you cannot evenly divide by 2. When you divide an odd number by 2, you get a left over value, known as the remainder.

PRIME NUMBERS

- A prime number is a positive number that you can only evenly divide by itself and the number 1.

Note: The number 1 is not considered a prime number since it can only be evenly divided by one number. The number 2 is the smallest prime number.

COMPOSITE NUMBERS

- A composite number is a number that you can evenly divide by itself, the number 1 and one or more other numbers.

- The numbers you can evenly divide into another number are called factors. For example, the factors for the number 12 are 1, 2, 3, 4 and 6.

CONTINUED

Classifying Numbers
continued

Think of these number classification groups as if they were clubs. Some clubs are relatively small because very few people meet their criteria for membership. Other clubs are huge due to the fact that they will let just about anybody join in. The prime number club, for example, is relatively exclusive and far smaller than the real number club, which will accept just about any number that walks in off the street.

It's important to remember, however, that just because a number belongs to an elite club, that does not keep it from belonging to other less-exclusive groups. In fact, as you are likely beginning to notice, most numbers usually meet the criteria to belong to a variety of these number clubs. The number 5 for example is a prime number, but it is also a positive number, a whole number and a real number.

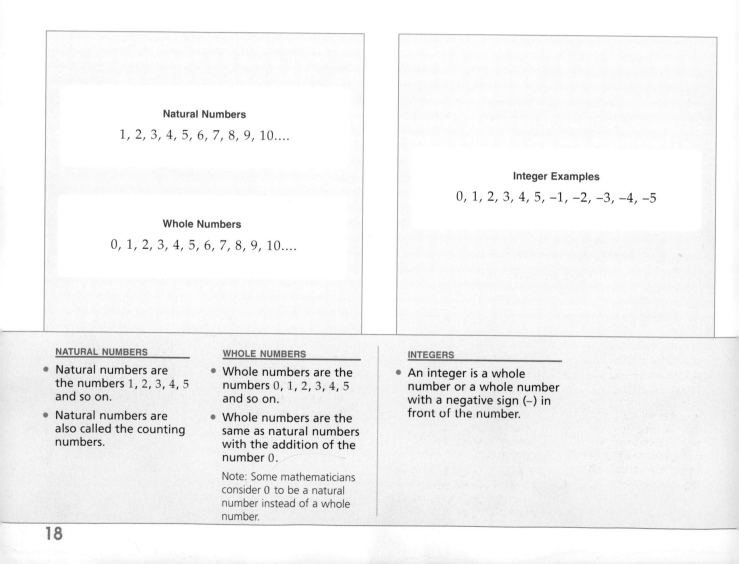

Natural Numbers

1, 2, 3, 4, 5, 6, 7, 8, 9, 10....

Whole Numbers

0, 1, 2, 3, 4, 5, 6, 7, 8, 9, 10....

Integer Examples

0, 1, 2, 3, 4, 5, –1, –2, –3, –4, –5

NATURAL NUMBERS

- Natural numbers are the numbers 1, 2, 3, 4, 5 and so on.
- Natural numbers are also called the counting numbers.

WHOLE NUMBERS

- Whole numbers are the numbers 0, 1, 2, 3, 4, 5 and so on.
- Whole numbers are the same as natural numbers with the addition of the number 0.

  Note: Some mathematicians consider 0 to be a natural number instead of a whole number.

INTEGERS

- An integer is a whole number or a whole number with a negative sign (–) in front of the number.

Tip

Why do some numbers have a line drawn over the decimal places?

A line drawn over decimal places indicates a repeating pattern. For instance, $\frac{1}{6}$ equals 0.166666666666..., with the 6s repeating on forever. To make the number easier to handle, it is expressed as follows:

$$0.1\overline{6}$$

Practice

Classify each of the following numbers into the natural, whole, integer, rational, irrational and real number groups. You can check your answers on page 250.

1) −1

2) $\sqrt{34}$

3) −18

4) π

5) $\frac{1}{3}$

6) 0

Rational Number Examples

$5, -5, 7, 55, \frac{1}{2}, \frac{1}{3}, \frac{3}{4}$

$9.25, 5.23, 0.2424242424...$

Irrational Number Examples

π (pi), $\sqrt{2}$ (square root of 2),
$\sqrt{3}$ (square root of 3)

Real Number Examples

$3, -3, 0, -55, \frac{2}{3}, \frac{1}{6}, 7.35, 4.1313..., \sqrt{2}, \pi$

RATIONAL NUMBERS

- Rational numbers include integers and fractions. A rational number is also a number that you can write with decimal values that either end or have a pattern that repeats forever.

IRRATIONAL NUMBERS

- Irrational numbers do not include integers or fractions. An irrational number has decimal values that continue forever without a pattern.
- The most well-known irrational number is pi (π), which is equal to 3.1415926...

REAL NUMBERS

- Real numbers include natural numbers, whole numbers, integers, rational numbers and irrational numbers.
- Real numbers can include fractions as well as numbers with or without decimal places.

Add and Subtract Numbers

At the very base of math are the functions of addition and subtraction. These functions come almost naturally to everyone. For example, if you are going out to a movie with two of your friends, you are going to buy three tickets. You have just added 1 + 2 to get an answer of 3.

Algebra throws one of its many curveballs with the introduction of negative numbers. Not to worry though, if you are like most people, you have already added negative numbers and not even realized it. Just think of negative numbers in terms of money. If you owe $10 to a friend

and then borrow $5 more, not only do you have a trusting friend who is short $15, but you have added negative numbers. In mathspeak, your financial statement would look like this: $-10 + -5 = -15$.

Working with negative numbers is not much different from simple addition and subtraction. The answer to an addition problem is still called the "sum" and the answer to a subtraction problem is still called the "difference" regardless of whether the answer ends up being a positive or negative number.

Adding Positive Numbers and Variables

$$2 + 3 = 5$$

$$50 + 30 = 80$$

$$8 + 10 + 12 = 30$$

$$x + 4$$

$$x + y + z$$

Adding Negative Numbers

$$(-2) + (-3) = -(2 + 3) = -5$$

$$(-5) + (-4) = -(5 + 4) = -9$$

$$(-6) + (-8) = -(6 + 8) = -14$$

$$3 + (-2) = 3 - 2 = 1$$

$$(-4) + 2 = -(4 - 2) = -2$$

1 When you see a plus sign (+) between two numbers, you add the numbers together. For example, 2 + 3 equals 5.

- You can also add variables and numbers together. A variable is a letter, such as x or y, which represents an unknown number.

- When adding two positive numbers, the answer will always be a positive number.

1 To add two negative numbers, add the numbers as if they were positive numbers and then place a negative sign (–) in front of the answer.

- When adding two negative numbers, the answer will always be a negative number.

2 To add a positive and a negative number, pretend both numbers are positive and subtract the smaller number from the larger number. Place the sign (+ or –) of the larger number in front of the answer.

- When adding a positive and a negative number, the answer could be a positive or negative number.

Tip

How can I simplify problems that contain signs which are side-by-side?

When you see two positive (+) signs or two negative (–) signs beside one another, you can replace the signs with one plus (+) sign.

$3 + (+5) - 3 + 5 = 8$

$6 - (-3) = 6 + 3 = 9$

When you see two different signs (+ or –) beside one another, you can replace the signs with one minus (–) sign.

$7 + (-3) = 7 - 3 = 4$

$9 - (+2) = 9 - 2 = 7$

Practice

$a + b = c$

Solve the following problems by applying the rules you have learned for adding and subtracting positive and negative numbers. You can check your answers on page 250.

1) $2 + 5 + 7$

2) $3 + (-5)$

3) $-7 + 10$

4) $-2 + (-4) + 11$

5) $1 - (-7)$

6) $6 - 3 - 9$

Subtracting Positive Numbers and Variables

$5 - 3 = 2$

$20 - 5 - 3 = 12$

$30 - 50 - -20$

$15 - x$

$x - y - z$

Subtracting Negative Numbers

$3 - (-2) = 3 + (+2) = 5$

$8 - (-5) = 8 + (+5) = 13$

$-5 - (-3) = -5 + (+3) = -2$

$-4 - (-7) = -4 + (+7) = 3$

$-6 - 3 = -6 + (-3) = -9$

---

1 When you see the minus sign (–) between two numbers, you subtract the second number from the first number. For example, $5 - 3$ equals 2.

- When subtracting a larger positive number from a smaller positive number, the answer will be a negative number. For example, $30 - 50$ equals –20.

- You can also subtract variables.

- To make a subtraction problem easier to solve when working with negative numbers, you can change a subtraction problem into an addition problem.

1 In a subtraction problem, change the minus sign (–) to a plus sign (+) and then change the sign of the number that follows from positive (+) to negative (–) or from negative (–) to positive (+). For example, $3 - (-2)$ equals $3 + (+2)$.

2 You can now solve the problem by adding the numbers.

21

Multiply and Divide Numbers

Algebra forces you to sharpen your multiplication and division skills. Aside from multiplying and dividing basic numbers, algebra will throw negative numbers and a bunch of letters called "variables" at you, which represent unknown numbers. As you will see, you can just apply the simple operations you already know.

The notation for multiplication may be a little different than you are used to. The variable "x" is often used in algebra problems, so the multiplication sign "×" you are accustomed to is often replaced by symbols such as "*" and "·" to avoid confusion. Sometimes the multiplication symbol is left out completely and you will need to multiply a number shown next to a variable or parenthesis. In a multiplication problem, the values being multiplied are known as the factors and the answer is the product.

In a division problem, the number being divided is called the dividend, the number that is dividing is the divisor and the answer is called the quotient. You will also find out that a horizontal fraction bar is not an apparatus for gymnastics, but just another way to denote division.

Multiplying Numbers and Variables

$$2 * 2 = 4$$

$$3 \cdot 5 = 15$$

$$6 \times 3 = 18$$

$$4(6) = 24$$

$$5x = 5 \times x$$

$$xy = x \times y$$

Dividing Numbers and Variables

$$6 \div 2 = 3$$

$$10/5 = 2$$

$$\frac{8}{2} = 4$$

$$x \div 3$$

$$x \div y$$

- When you see the multiplication sign (*, · or ×) between two numbers, you multiply the numbers together. For example, 2×2 equals 4.

- When you see a number beside a set of parentheses ( ), you multiply the numbers together. For example, 4(6) equals 4×6 equals 24.

- You can also multiply variables. A variable is a letter, such as x or y, which represents an unknown number.

- If a symbol does not appear between a number and variable or two variables, you multiply the number and variable or variables together. For example, $5x$ equals $5 \times x$ and xy equals $x \times y$.

- When you see the division sign (÷ or /) between two numbers, you divide one number by the other number. For example, $6 \div 2$ equals 3.

- Horizontal fraction bars (–) are also used to indicate division. For example, $\frac{8}{2}$ equals $8 \div 2$.

- You can also divide variables.

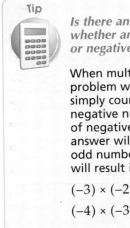

Tip

Is there an easy way to determine whether an answer will be positive or negative?

When multiplying or dividing a problem with a lot of numbers, simply count the number of negative numbers. An even number of negative numbers means the answer will be positive, while an odd number of negative numbers will result in a negative answer.

$(-3) \times (-2) \times (-5) \times (-1) = 30$

$(-4) \times (-3) \times (-2) = -24$

Practice

Solve the following problems by applying the rules you have learned for multiplying and dividing positive and negative numbers and variables. You can check your answers on page 250.

1) $(-1) \times (-5)$

2) $3 \cdot 2$

3) $\dfrac{10}{2}$

4) $x(-2)$

5) $(4)(3)$

6) $(4) \div (-2)$

Multiplying or Dividing Numbers with the Same Sign
$2 \times 5 = 10$
$(-3) \times (-6) = 18$
$a \times b = ab$
$(-a) \times (-b) = ab$
$28 \div 4 - 7$
$(-18) \div (-9) = 2$
$a \div b = a \div b$
$(-a) \div (-b) = a \div b$

Multiplying or Dividing Numbers with Different Signs
$(+3) \times (-2) = -6$
$(-4) \times (+6) = 24$
$(+a) \times (-b) = -ab$
$(-a) \times (+b) = -ab$
$24 \div (-8) - -3$
$(-28) \div (+4) = -7$
$(+a) \div (-b) = -(a \div b)$
$(-a) \div (+b) = -(a \div b)$

- If the two numbers or variables you are multiplying or dividing have the same sign (+ or –), the result will always be positive.

- For example, 2×5 and $(-2) \times (-5)$ both equal 10.

- If the two numbers or variables you are multiplying or dividing have different signs (+ and –), the result will always be negative.

- For example, $(+3) \times (-2)$ and $(-3) \times (+2)$ both equal –6.

Absolute Values

Absolute value is a basic concept you will often encounter in algebra. Absolute values tell you how far a given value is from 0. For example, while 6 and –6 are on opposite sides of 0 on the number line, both are 6 moves away from 0. Therefore, the absolute value of both 6 and –6 is 6. Absolute values are always positive.

Absolute value symbols (| |) appear around any number for which you are trying to find the absolute value. For example, the absolute value of –7 is written as $|-7|$.

Calculating the absolute value of a number is simple. All you have to do is remove the negative sign, if there is one. For example, the absolute value of –4 is 4. Even though you convert negative numbers to positive numbers, absolute values are not always opposite values. The absolute value of a positive value remains positive. Just remember that absolute values are eternal optimists—positive no matter what.

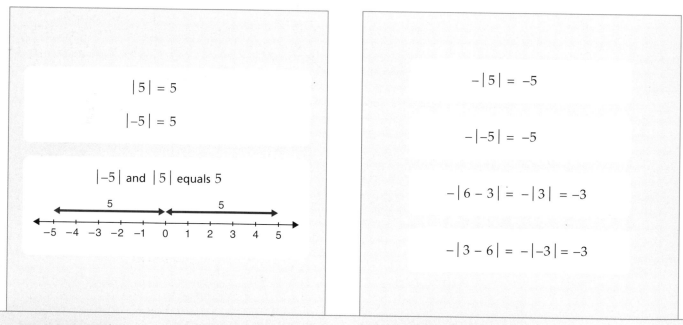

$$|5| = 5$$

$$|-5| = 5$$

$|-5|$ and $|5|$ equals 5

$$-|5| = -5$$

$$-|-5| = -5$$

$$-|6 - 3| = -|3| = -3$$

$$-|3 - 6| = -|-3| = -3$$

- The absolute value of a number is always the positive value of the number, whether the number is positive or negative.
- For example, the absolute value of 5 and –5 is 5.

- An absolute value is indicated by two thin vertical lines (| |) around a number or numbers.
- An absolute value indicates how far a number is from zero, regardless of whether the number is positive or negative. For example, 5 and –5 have the same absolute value since they are both 5 units from zero.

- If a negative sign (–) appears in front of absolute value symbols (| |), after you determine the absolute value of a number, you will need to place a negative sign (–) in front of the number. For example, $-|5|$ equals –5.

Grouping Symbols

Grouping symbols are your friends. When you are confronted with a long, mind-boggling algebra problem, the grouping symbols—parentheses ( ), brackets [ ] and braces { }—will guide you. Grouping symbols are used to show you where you should begin as you work to simplify a problem. Whatever operations appear inside the symbols are considered grouped together and should be the first thing you tackle.

No set of grouping symbols takes prominence over another, though you will probably see parentheses most often.

You will find that grouping symbols can be used separately in a problem or within one another. When sets of grouping symbols appear within one another, you should begin with the innermost set and work your way outward.

About Grouping Symbols
$8 - (5 \times 2)$ $= 8 - 10$ $= -2$
$20 \div \{6 - 2\}$ $= 20 \div 4$ $= 5$
$16 - [2 \times (3 + 4)]$ $= 16 - [2 \times 7]$ $= 16 - 14$ $= 2$

Working with Grouping Symbols and Variables
$100 + (a \times b)$
$x \div \{y - z\}$
$20 - [x \times (y + 6)]$
$(x + y)(6 - z)$
$6 \times (x - y - 8)$

- You will see several types of grouping symbols, including parentheses ( ), brackets [ ] and braces { }.
- You should always work with numbers inside grouping symbols first.

- If more than one set of grouping symbols appear in a problem, work with the numbers in the innermost set of grouping symbols first. Then work your way to the outside of the problem.

- As with numbers, you will also see grouping symbols used with variables. A variable is a letter, such as x or y, which represents an unknown number.

Order of Operations

Now that you know how to work with negative numbers and variables, you may feel ready to take on more complex problems involving these concepts. Not so fast. Tackling a problem using the left-to-right strategy you use when reading may seem logical but can give you flawed results.

Rebels that they are, mathematicians have decided that they do not always want to work from left to right. To avoid confusion they came up with the Order of Operations—working first with parentheses, then exponents, then multiplication and division, and finally with addition and subtraction. Following the Order of Operations ensures that anyone working on a problem will arrive at the same answer every time.

Think of solving a math problem using the Order of Operations like building a pizza. Try putting the toppings and cheese down on the tray before the dough and you won't end up with a pizza, just a headache for the person left to wash the dishes.

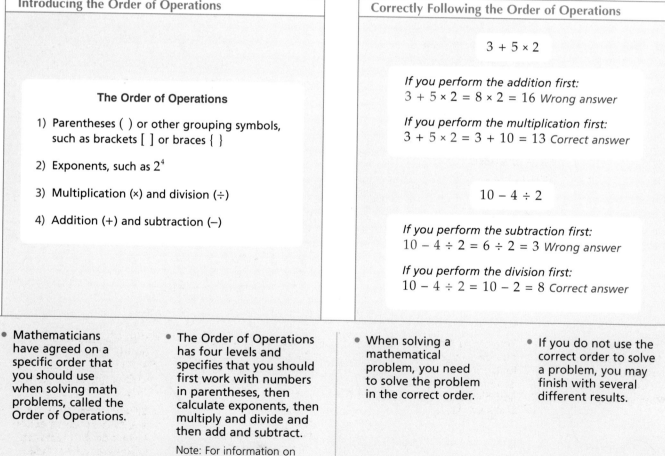

Introducing the Order of Operations

The Order of Operations

1) Parentheses ( ) or other grouping symbols, such as brackets [ ] or braces { }

2) Exponents, such as 2^4

3) Multiplication (×) and division (÷)

4) Addition (+) and subtraction (−)

Correctly Following the Order of Operations

$$3 + 5 \times 2$$

If you perform the addition first:
$3 + 5 \times 2 = 8 \times 2 = 16$ *Wrong answer*

If you perform the multiplication first:
$3 + 5 \times 2 = 3 + 10 = 13$ *Correct answer*

$$10 - 4 \div 2$$

If you perform the subtraction first:
$10 - 4 \div 2 = 6 \div 2 = 3$ *Wrong answer*

If you perform the division first:
$10 - 4 \div 2 = 10 - 2 = 8$ *Correct answer*

- Mathematicians have agreed on a specific order that you should use when solving math problems, called the Order of Operations.

- The Order of Operations has four levels and specifies that you should first work with numbers in parentheses, then calculate exponents, then multiply and divide and then add and subtract.

  Note: For information on exponents, see page 54.

- When solving a mathematical problem, you need to solve the problem in the correct order.

- If you do not use the correct order to solve a problem, you may finish with several different results.

Tip

Is there an easy way to remember the Order of Operations?

Using the first letter from each operation, you can sum up the Order of Operations in a simple sentence, such as "**P**lease **E**xcuse **M**y **D**ear **A**unt **S**ally."

Parentheses
Exponents
Multiplication/**D**ivision
Addition/**S**ubtraction

Similarly, some people prefer to condense the Order of Operations into a single word, such as **PEDMAS**, or **BEDMAS**, where "B" stands for "brackets."

Practice

$a + b = c$

Solve the following problems by applying the Order of Operations. You can check your answers on page 250.

1) $3 - (2 - 5)$

2) $(4 - 2) \times 3 + (1 - 3)^3$

3) $\dfrac{(3 - 5)}{(9 - 7)}$

4) $2 - 3 \times 5$

5) $\dfrac{4}{2} \times 3$

6) $2^{(4 - 2)}$

Working with the Order of Operations

$8 + 2 \times 3 - 6 \div 2$ Multiply and divide first.
$= 8 + 6 - 3$ Then add and subtract.
$= 11$

$(5 - 3)^2 - 2 \times 4$ Work with numbers in parentheses first.
$= (2)^2 - 2 \times 4$ Then calculate the exponent.
$= 4 - 2 \times 4$ Then multiply.
$= 4 - 8$ Then subtract.
$= -4$

$4 + 12 \div (5 - 3)$ Work with numbers in parentheses first.
$= 4 + 12 \div (2)$ Then divide.
$= 4 + 6$ Then add.
$= 10$

Working with Operations on the Same Level

$20 - [2 \times (3 + 5)]$
$= 20 - [2 \times 8]$
$= 20 - 16$
$= 4$

$5 - 3 + 6$
$= 2 + 6$
$= 8$

- The above examples demonstrate the correct order of operations when solving problems.

- Remember to work with numbers in parentheses first, then calculate exponents, then multiply and divide and then add and subtract.

- When working on the parentheses level and one set of parentheses appears within another, start with the innermost set and work your way to the outside of the problem.

- When working on the multiplication and division level, multiply or divide, whichever comes first, from left to right.

- When working on the addition and subtraction level, add and subtract, whichever comes first, from left to right.

The Commutative Property

Have you ever seen a game of three-card monte? It is that scam where a hustler reveals one of three cards and bets that the player cannot find the card after a quick shuffle. No matter how the cards are mixed, the result is always the same—the player cannot find the card.

The commutative property is similar, but without the con artists. The commutative property states that when calculating addition or multiplication problems, you will arrive at the same answer no matter what order the terms are arranged in.

For example, both 3 + 4 + 5 and 5 + 3 + 4 equal 12. The commutative property does not apply to subtraction and division problems, which must be carried out in the order that they were given. You will find that the commutative property is a great way to reorganize cluttered equations.

Be careful not to confuse the commutative property with the associative property, which deals with how the terms are grouped. You can read more about the associative property on page 29.

Numbers	Variables
$2 + 3 = 5$ $3 + 2 = 5$	$x + y = y + x$ $x \times y = y \times x$
$2 \times 3 = 6$ $3 \times 2 = 6$	$a + b + c$ $= c + a + b$ $= b + c + a$
$2 \times 3 \times 4 = 24$ $3 \times 4 \times 2 = 24$ $4 \times 2 \times 3 = 24$	$a \times b \times c$ $= c \times a \times b$ $= b \times c \times a$

- When you add or multiply two or more numbers together, the order in which you add or multiply the numbers will not change the result. This is known as the commutative property of addition and multiplication.

- For example, whether you add 2 + 3 or 3 + 2, the answer is always 5.

- As with numbers, when you add or multiply two or more variables, the order in which you add or multiply the variables will not change the result. A variable is a letter, such as x or y, which represents an unknown number.

- For example, $x + y$ is equal to $y + x$.

The Associative Property

The associative property states that when calculating addition or multiplication problems, you will arrive at the same answer no matter what order the operations are performed in. For example, both $(2 \times 5) \times 6$ and $2 \times (5 \times 6)$ equal 60.

The following scenario illustrates the associative property. Ruben, Nicky and Beverly each decide to give five cans of food to the local food bank. Ruben could take his five cans along with Nicky's five cans and meet Beverly with her five cans at the food bank, or Beverly could pick up Nicky's donation and meet Ruben at the food bank. No matter how the donations arrive at the food bank, the total donation will be 15 cans. That is the associative property in a nutshell.

The associative property does not apply to subtraction and division. The order in which you subtract or divide numbers, as defined by grouping symbols, will change the results.

The associative property is handy when you need to reorganize cluttered equations. Be careful, though, not to confuse the associative property with the commutative property, which deals with the order in which terms are arranged. You can read more about the commutative property on page 28.

Numbers
$(2 + 4) + 5 = 6 + 5 = 11$ $2 + (4 + 5) = 2 + 9 = 11$ $(3 + 2) + 7 = 5 + 7 = 12$ $3 + (2 + 7) = 3 + 9 = 12$ $(2 \times 3) \times 4 = 6 \times 4 = 24$ $2 \times (3 \times 4) = 2 \times 12 = 24$

Variables
$(a + b) + c = a + (b + c)$ $(4 + a) + b = 4 + (a + b)$ $a \times (b \times c) = (a \times b) \times c$ $x \times (2 \times y) = (x \times 2) \times y$

- When you either add or multiply more than two numbers together, the order in which you either add or multiply the numbers as defined by the grouping symbols, such as ( ), will not change the result. This is known as the associative property of addition and multiplication.

Note: When you see grouping symbols, such as parentheses ( ), you should always work with the numbers inside the grouping symbols first. For more information on grouping symbols, see page 25.

- As with numbers, when you add or multiply more than two variables together, the order in which you add or multiply the variables as defined by the grouping symbols, such as ( ), will not change the result.

- A variable is a letter, such as x or y, which represents an unknown number.

The Distributive Property

Since childhood, we have been taught that by sharing we can break down barriers. In a case of life imitating math, you can see the sharing principle at work in algebra through something called the distributive property.

Using the distributive property, you can eliminate a set of grouping symbols, such as parentheses ( ), by "sharing," or multiplying, each number and variable within the parentheses by a number or variable outside of the parentheses. For example, if you have $5(4 + a)$, you multiply the number and variable within the parentheses by 5 to evenly

distribute the 5 to everything inside the parentheses. In this example, $5(4 + a) = (5 \times 4) + (5 \times a) = 20 + 5a$.

Distributing a positive number or variable is straightforward, but you have to take care when distributing negative numbers or variables. When you distribute a negative number or variable, all of the terms inside the parentheses change signs from positive to negative and/or from negative to positive. For example, $-1(2 + x)$ becomes $-2 - x$.

Positive Numbers

$$2(6 + 3) = (2 \times 6) + (2 \times 3)$$
$$= 12 + 6$$

$$6(4 + 5 + 9) = (6 \times 4) + (6 \times 5) + (6 \times 9)$$
$$= 24 + 30 + 54$$

$$4(7 - 5) = (4 \times 7) - (4 \times 5)$$
$$= 28 - 20$$

Negative Numbers

$$-3(2 + 4) = (-3 \times 2) + (-3 \times 4)$$
$$= (-6) + (-12)$$

$$-2(4 - 5) = (-2 \times 4) - (-2 \times 5)$$
$$= (-8) - (-10)$$
$$= -8 + 10$$

$$-(3 + 2) = (-1 \times 3) + (-1 \times 2)$$
$$= (-3) + (-2)$$

1 To remove the parentheses ( ) that surround numbers you need to add or subtract, you can multiply the number outside the parentheses by each number inside the parentheses.

1 To remove the parentheses ( ) that surround numbers you need to add or subtract, you can multiply the negative number outside the parentheses by each number inside the parentheses.

● If you see a negative sign (−) by itself outside the parentheses, assume the number in front of the parentheses is −1. For example $-(3 + 2)$ equals $-1(3 + 2)$.

Note: When you multiply two numbers with different signs (+ or −), the result will be a negative number. For example, -3×2 equals −6.

Tip

How can I make problems with several positive (+) and negative (–) signs less complicated?

If a problem contains positive or negative signs which are side by side, you can simplify the problem a little bit. When you find two positive signs or two negative signs next to one another, you can replace the signs with a single plus (+) sign.

$3 + (+5) = 3 + 5$
$6 - (-3) = 6 + 3$

If you find two opposite signs (+ or –) side by side, you can replace the signs with a single minus (–) sign.

$7 + (-3) = 7 - 3$
$9 - (+2) = 9 - 2$

Practice

a+b=c

Simplify the following expressions by using the distributive property. You can check your answers on page 250.

1) $2(3 - 5)$

2) $3(7 - 4)$

3) $5(1 + 2)$

4) $-1(10 + 4)$

5) $-2(3 - 2)$

6) $-5(-1 - 4)$

Variables

$$3(2 + x) = (3 \times 2) + (3 \times x)$$
$$= 6 + 3x$$

$$-4(x - y) = (-4 \times x) - (-4 \times y)$$
$$= -4x - (-4y)$$
$$= -4x + 4y$$

$$a(a + b + c) = (a \times a) + (a \times b) + (a \times c)$$
$$= a^2 + ab + ac$$

Variables with Exponents

$$3(x^2 + y^2) = (3 \times x^2) + (3 \times y^2)$$
$$= 3x^2 + 3y^2$$

$$x^2(7 - x^2) = (x^2 \times 7) - (x^2 \times x^2)$$
$$= 7x^2 - x^{2+2}$$
$$= 7x^2 - x^4$$

$$-x^3(x^2 + y^4) = (-x^3 \times x^2) + (-x^3 \times y^4)$$
$$= (-x^{3+2}) + (-x^3y^4)$$
$$= -x^5 - x^3y^4$$

1 To remove the parentheses ( ) that surround numbers and variables you need to add or subtract, you can multiply the number or variable outside the parentheses by each number and variable inside the parentheses.

● A variable is a letter, such as x or y, which represents an unknown number.

Note: When you multiply a variable by itself, you can use exponents to simplify the problem. For example, $a \times a$ equals a^2. For more information on exponents, see page 54.

1 To remove the parentheses ( ) that surround numbers and variables with exponents you need to add or subtract, you can multiply the number or variable outside the parentheses by each number and variable inside the parentheses.

Note: When multiplying numbers or variables with exponents that have the same base, you can add the exponents. For example, $x^3 \times x^2$ equals x^{3+2}, which equals x^5.

The Inverse Property

The inverse property outlines the two types of opposites in algebra—the additive inverse and the multiplicative inverse.

The additive inverse property simply states that every number has an inverse number with the opposite sign. For example, the additive inverse of 12 is –12. When you add two opposites together, the numbers will effectively cancel each other out and bring you to 0.

The multiplicative inverse property states that every number, except for zero, has an opposite reciprocal number. When a number is multiplied by its reciprocal, the answer will always be 1. A reciprocal is found by simply flipping a fraction over. For example, to find the reciprocal of 2, rewrite it as the fraction $\frac{2}{1}$—keeping in mind that all positive and negative whole numbers have an invisible denominator, or bottom part of the fraction, of 1—and flip the fraction over so that you have $\frac{1}{2}$. When you multiply 2 by $\frac{1}{2}$, you arrive at 1.

The relationships between numbers and their opposites may not seem like a big deal now, but in algebra you will be constantly unraveling complicated problems. The additive and multiplicative inverse properties will be a couple of your main tools.

Additive Inverse Property

Additive Inverse
$$3 + (-3) = 0$$

$$10 + (-10) = 0$$

$$7 + (-7) = 0$$

$$-5 + 5 = 0$$

$$-30 + 30 = 0$$

$\Rightarrow$

Additive Inverse
$$a + (-a) = 0$$

$$b + (-b) = 0$$

$$-b + b = 0$$

$$-x + x = 0$$

$$-y + y = 0$$

- When you add a number to its opposite, called the additive inverse, the result will be 0.

- The additive inverse of a number is the same number, but with a different sign (+ or –). For example, the additive inverse of 3 is -3.

- As with numbers, when you add a variable to its opposite, or additive inverse, the result will be 0.

- A variable is a letter, such as x or y, which represents an unknown number.

Evaluate an Expression

As you may have noticed, mathematical statements do not always use numbers exclusively. Often the statements contain one or more variables. If you know the value of each variable, you can solve the problem—also known as evaluating an expression.

To evaluate an expression, you replace the variables in the expression with their numerical values, otherwise known as plugging numbers into an expression. For the best results, you should surround the values you plug into an expression with parentheses so that positive and negative numbers don't get mixed up with the mathematical operations of the problem. For example, plugging the value of $x = -4$ into the equation $y + 4x$ would be open to errors if it was written as $y + 4 - 4$. Using parentheses and writing the expression as $y + 4(-4)$ removes any ambiguity. Once you have replaced the variables in an expression, you simplify the expression using the normal rules—parentheses first, then exponents, then multiplication and division, and then addition and subtraction.

You will frequently evaluate an expression to check your answers after you have solved for a variable.

Evaluate the expression $x + y + z$
when $x = 2$, $y = 3$ and $z = 4$

$x + y + z$

$= (2) + (3) + (4) = 9$

Evaluate the expression $20 \div (x - y)$
when $x = 9$ and $y = 4$

$20 \div (x - y)$

$= 20 \div (9 - 4) = 20 \div 5 = 4$

Evaluate the expression $2(x + y)$
when $x = 5$ and $y = 8$

$2(x + y)$

$= 2(5 + 8) = 2(13) = 2 \times 13 = 26$

Evaluate the expression $x^2 + (y - 4)$
when $x = 3$ and $y = 6$

$x^2 + (y - 4)$

$= (3)^2 + (6 - 4) = 9 + 2 = 11$

- An expression consists of numbers, variables or a combination of the two, which are connected by signs, such as $+$, $-$, $\times$ or $\div$.

  Note: A variable is a letter, such as x or y, which represents an unknown number.

- When you are asked to evaluate an expression, you will need to find the answer to the problem.

1 To evaluate an expression, replace the variables with the given number for each variable and then simplify the expression.

- Above are more examples of evaluating expressions.

  Note: When a sign does not appear between a number and variable, such as $3x$, you multiply the number and variable together.

Note: For information on exponents, such as x^2, see page 54.

Add and Subtract Variables

You can use your addition and subtraction skills to simplify expressions that use variables even before you have figured out the values of the variables. For example, when simplifying the expression $y + y + y$, we know there are three ys. The three ys in the expression can be written as $3 \times y$, or just $3y$ for short. Even though you may not know the value of a variable, you can combine the same variables in an expression because they represent the same values.

Beware of different variables though. Different variables are like oil and water when it comes to addition and subtraction—they do not mix. $5a - 2b$ does not equal $3ab$. Similarly, terms with different exponents of the same variable, like x and x^2 are not like terms and cannot be added or subtracted. You must leave the expression the way it was originally written until you can figure out the values for one or more variables.

Understanding Like Terms

Like Terms

$x, 2x, 3x, 4x, 5x$

$xy, 2xy, 3xy, 4xy, 5xy$

$abc, 2abc, 3abc, 4abc, 5abc$

Unlike Terms

$2x, 6y, 5z$

$5, x, 7y$

$3ab, 5ac, 8ad$

- Each item you add or subtract is called a term. For example, $2x$, y and $5ac$ are all terms.
- You can only add and subtract like terms, which are terms with the same variable(s). A variable is a letter, such as x or y, which represents an unknown number. For example, x, $2x$ and $3x$ are like terms.

- You cannot add and subtract unlike terms, which are terms with different variables. For example, x, y and z are unlike terms. Numbers and variables are also unlike terms. For example, 5 and $4x$ are unlike terms.

Add Variables

$2x + 3x = 5x$

$5y + y = 6y$

$4xy + 7xy = 11xy$

$7a + 2a + 6 = 9a + 6$

$3a + 6a + 4b + 2b = 9a + 6b$

1 To add like terms, add the numbers in front of the variables and place the resulting number in front of the shared variable. For example, $2x + 3x$ equals $5x$.

Note: The number in front of a variable is called the coefficient. For example, in the expression $2x$, 2 is the coefficient and x is the variable.

- When a number does not appear in front of a variable, assume the number is one. For example, $a + a$ is equal to $1a + 1a$.

Tip

How can I make long expressions containing exponents easier to follow?

In your answer, you should list the exponents from highest to lowest. Not only will listing exponents in order make the expression easier to understand, it is also considered good form. For example, instead of writing $7a - 5a^3 + 3a^4 - 3a^2$, write the expression as $3a^4 - 5a^3 - 3a^2 + 7a$.

Practice

$a + b = c$

Simplify the following expressions by adding and subtracting variables. You can check your answers on page 251.

1) $3x + 2x^3 + 2x + x^2 - 5x^3$

2) $4 - x + 3 + 3x$

3) $x^5 + 7x - 4x + x^4 + x^5$

4) $x - 2x + 3x + x - x$

5) $5 + x^2 - 3 - x^2 - 2$

6) $x + x^7 + 5 - 4x - x^7$

Subtract Variables

$$8x - 6x = 2x$$

$$6y - y = 5y$$

$$5xy \quad 7xy = 2xy$$

$$9a - 5a - 2 = 4a - 2$$

$$20a - 10a - 7b - 2b = 10a - 9b$$

Add or Subtract Variables with Exponents

$$2x^2 + 3x^2 = 5x^2$$

$$6y^3 + y^3 = 7y^3$$

$$7x^2y^2 - 4x^2y^2 = 3x^2y^2$$

$$5a^2 - 2a^2 + 6 = 3a^2 + 6$$

$$4a^2 + 2a^2 - 6b^3 + 5b^3 = 6a^2 - b^3$$

---

1 To subtract like terms, subtract the numbers in front of the variables and place the resulting number in front of the shared variable. For example, $8x - 6x$ equals $2x$.

- When a number does not appear in front of a variable, assume the number is one. For example, $a - b$ is equal to $1a - 1b$.

- You can add and subtract variables with the same exponents the same way you would add and subtract variables without exponents.

  Note: For information on exponents, see page 54.

1 To add or subtract terms with the same variable and the same exponent, add or subtract the numbers in front of the variables and place the resulting number in front of the shared variable. For example, $2x^2 + 3x^2$ equals $5x^2$.

Multiply and Divide Variables

Almost anything goes when you are multiplying and dividing variables. There are ways to work out just about any combination of numbers and variables. For instance, if you want to multiply a number by a variable, $3 \times y$ for example, just write $3y$ and the multiplication symbol is implied. If you are dividing 3 by y, simply write the expression as a fraction $\frac{3}{y}$.

You want to multiply or divide two variables? Piece of cake. Unlike when adding and subtracting, dissimilar variables are no big deal.

For instance, $3a \times 7b$ multiplies everything together to come out $21ab$ and $a \div b$ can be easily expressed as the fraction $\frac{a}{b}$.

For your last trick, you can even multiply and divide variables with exponents. You have to keep on your toes, but you can do it with relative ease. Just remember that the rules for multiplying and dividing exponents still apply. When you are working with two or more of the same variable, you just add and subtract the exponents. For example, the problem $5c^4 \times 3c^3$ equals $5 \times 3c^{4+3}$, which works out to $15c^7$.

Multiplying a Variable by a Number

$$8x \times 4 = 32x$$

$$6 \times 3y = 18y$$

$$x \times 2 = 1x \times 2 = 2x$$

$$5 \times a = 5 \times 1a = 5a$$

$$7b \times 2 = 14b$$

Multiplying Two or More Variables

$$3x \times 2y = 6xy$$

$$5xy \times 4xy = 20x^2y^2$$

$$6a \times 2a \times 2b = 24a^2b$$

$$5x \times 2x \times 3x = 30x^3$$

$$2a^3b^5 \times 3a^2b^4 = 6a^{3+2}b^{5+4} = 6a^5b^9$$

1 To multiply a variable by a number, multiply the number in front of the variable, called the coefficient, by the number. A variable is a letter, such as x or y, which represents an unknown number.

Note: When a number does not appear in front of a variable, assume the number is one. For example, x equals $1x$.

2 Place the resulting number in front of the variable.

- For example, $8x \times 4$ equals $32x$.

1 To multiply two or more variables together, multiply the numbers in front of the variables. Then multiply the variables together.

2 Place the resulting number in front of the resulting variables.

- For example, $3x \times 2y$ equals $6xy$.

Note: When multiplying variables with the same letter, such as x, you can add the exponents. If a variable does not have an exponent, assume the exponent is one. For example, $x \times x$ is the same as $x^1 \times x^1$, which equals x^2.

Practice

$a+b=c$

Solve the following problems by multiplying variables. You can check your answers on page 251.

1. $(3x^3)(4x^2)$

2. $(4y^5)(-2y^{-3})$

3. $3x^2 4y^3 2x^5 y^{-2}$

4. $(-x)(2y^{-1})(3y^3)(-x)$

5. $(-5z^2)(2z^{-5})$

6. $3x^2 y^3 \times 2x^{-3} y^4$

Practice

$a+b=c$

Solve the following problems by dividing variables. You can check your answers on page 251.

a) $\dfrac{10x^2}{5x}$

b) $15y^5 \div 3y^2$

c) $20x^2 y^2 \div 5x^2$

d) $\dfrac{6x^3 y^4}{2x y^2}$

e) $6x^2 \div 3x$

f) $4a^6 b^5 \div 2a^4 b^2$

Dividing a Variable by a Number

$$\frac{20x}{4} = 5 \times x = 5x$$

$$\frac{40ab^2}{10} = 4 \times ab^2 = 4ab^2$$

$$\frac{56}{7x} = \frac{8}{x}$$

$$\frac{72}{8x^3 y} = \frac{9}{x^3 y}$$

Dividing Variables

$$\frac{9x^5}{3x^3} = 3x^{5\,3} = 3x^2$$

$$\frac{3a^7}{4a^2} = \frac{3a^{7\,2}}{4} = \frac{3a^5}{4}$$

$$\frac{8x}{4x} = 2x^{1-1} = 2x^0 = 2 \times (1) = 2$$

$$\frac{50a^8 b^4}{10a^3 b^3} = 5a^{8-3} b^{4-3} = 5a^5 b^1 = 5a^5 b$$

---

1 To divide a variable by a number, divide the number in front of the variable by the number.

2 Multiply the resulting number by the variable.

- For example, $20x \div 4$ equals $5x$.

- To divide a number by a variable, divide the number by the number in front of the variable. The variable will remain in the denominator, or bottom part of the fraction.

1 To divide variables, divide the numbers in front of the variables. Then divide the variables.

Note: When dividing variables with the same letter, you can subtract the exponents. For example, $x^5 \div x^3$ equals x^{5-3}, which equals x^2. If a variable does not have an exponent, assume the exponent is one. For example, x equals x^1. A variable to the power of 0 equals one. For example, x^0 equals one.

2 Combine the resulting number with the resulting variables.

- For example, $9x^5 \div 3x^3$ equals $3x^{5-3}$, which equals $3x^2$.

Algebra Basics

Question 1. Perform the following operations.

a) $2 + 3 + 8$

b) $3 + (-7)$

c) $5 - 4 - 5$

d) $-12 - (-6)$

e) $2 - 7$

f) $3 - 4 + 2$

g) $-5 + 7 + (-2)$

h) $2 - (-3)$

Question 2. Perform the following operations.

a) $4 \div 2$

b) $12/4$

c) $(-8) \div 2$

d) $(-20)/(-5)$

e) $2(7)$

f) $3 \times (-5)$

g) $(-2)(-10)$

Question 3. Find the following absolute values.

a) $|3|$

b) $|-6|$

c) $|0|$

d) $-|3 - 7|$

e) $|12 - 3|$

Question 4. Evaluate the following expressions.

a) $2 - 3 \times 5$

b) $2 \times (4 - 2)$

c) $(5 - 3) \times (1 - 6)$

d) $3 - 20 \div 4$

e) $\dfrac{2 \times 10}{6 + 2 \times 7}$

Question 5. Find the multiplicative inverse of the following numbers.

a) $\dfrac{1}{4}$

b) $\dfrac{2}{3}$

c) -10

d) -1

e) $\dfrac{4}{5}$

Question 6. Simplify the following expressions as much as possible.

a) $x^2 + 2x^2 + x - 5x + 6 + 3 - 1$

b) $x^4 - x^2 + x^3 + x^4 + x^2 + 4x - 2 + x + 3$

c) $5x + 2y + 6xy - 7y - x - 3xy$

d) $4x \times 2x \times 3x$

e) $\dfrac{21x}{3}$

f) $12x^4 \div 3x^2$

Question 7. Perform the following multiplication and division operations.

a) $\dfrac{x \times x^3}{x^2}$

b) $\dfrac{4x^4}{2x^3}$

c) $\dfrac{x \times 6x^{10}}{(-2x^5)}$

d) $x(3 - 2y)$

e) $-5(3 - 7x)$

f) $4x(-3x^2 + 2x + 6y)$

You can check your
answers on page 266.

41

Chapter 2

If your experience with fractions amounts to digging for "quarters" in your pocket, this chapter will be a huge help. Soon you will be performing basic operations with fractions without a second thought. You will also learn about exponents, handy little numbers that have a big effect. Chapter 2 shows you how to deal with both fractions and exponents.

Working with Fractions and Exponents

In this Chapter...

Introduction to Fractions

If you have ever been shopping, you likely know all about fractions. Sales banners often use fractions because the values of fractions are easy to understand. Somehow, "$\frac{1}{2}$ off!" resonates with bargain hunters immediately.

Fractions are a tidy way of expressing part of a whole number or set of numbers and look neater than writing out a string of digits after a decimal point. The top number of the fraction, called the numerator, tells you how many units you have, while the bottom number of the fraction, called the denominator, tells you how many units are in a complete set.

You can view a fraction as a division problem that has been paused. If you finish the problem, you will end up with the decimal value of the fraction. For instance, $\frac{4}{5}$ is the same as $4 \div 5$—both equal 0.8.

Improper fractions are top-heavy fractions in which the numerator is larger than the denominator. An improper fraction can also be stated as a mixed number, which is a whole number followed by a fraction. For example, $\frac{3}{2}$ is an improper fraction that can be stated as the mixed number $1\frac{1}{2}$. Fractions with identical numerators and denominators equal 1 and are normally written as 1.

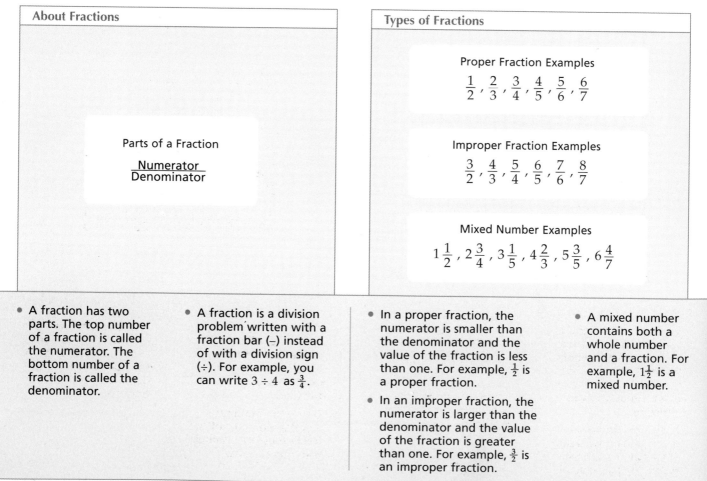

About Fractions

Parts of a Fraction

Numerator
Denominator

Types of Fractions

Proper Fraction Examples

$\frac{1}{2}, \frac{2}{3}, \frac{3}{4}, \frac{4}{5}, \frac{5}{6}, \frac{6}{7}$

Improper Fraction Examples

$\frac{3}{2}, \frac{4}{3}, \frac{5}{4}, \frac{6}{5}, \frac{7}{6}, \frac{8}{7}$

Mixed Number Examples

$1\frac{1}{2}, 2\frac{3}{4}, 3\frac{1}{5}, 4\frac{2}{3}, 5\frac{3}{5}, 6\frac{4}{7}$

- A fraction has two parts. The top number of a fraction is called the numerator. The bottom number of a fraction is called the denominator.

- A fraction is a division problem written with a fraction bar (–) instead of with a division sign (÷). For example, you can write $3 \div 4$ as $\frac{3}{4}$.

- In a proper fraction, the numerator is smaller than the denominator and the value of the fraction is less than one. For example, $\frac{1}{2}$ is a proper fraction.

- In an improper fraction, the numerator is larger than the denominator and the value of the fraction is greater than one. For example, $\frac{3}{2}$ is an improper fraction.

- A mixed number contains both a whole number and a fraction. For example, $1\frac{1}{2}$ is a mixed number.

Simplify Fractions

When working with a fraction, you can often simplify the fraction to place smaller numbers in the numerator, or top number, and denominator, or bottom number, of the fraction. Smaller numbers generally make fractions easier to work with. For example, the fraction $\frac{1}{4}$ is much easier to understand and work with than $\frac{12}{48}$, even though both fractions have the same value.

To simplify a fraction, you find a number that divides evenly into both the numerator and denominator of the fraction and then divide both parts of the fraction by that number. The most efficient method of simplifying a fraction is to determine the largest number that evenly divides into both the numerator and the denominator—known as the greatest common factor. When you divide both the numerator and denominator by the same number, the value of the fraction does not change.

Most teachers will not consider an answer complete if it contains a fraction that is not completely simplified. Keep in mind, though, that you will sometimes encounter fractions that cannot be simplified further. For example, the fraction $\frac{73}{91}$ cannot be simplified, since there are no numbers that divide evenly into both parts of the fraction.

Examples of Simplified Fractions

$$\frac{1}{2} , \frac{2}{3} , \frac{3}{4} , \frac{4}{5} , \frac{5}{6} , \frac{6}{7} , \frac{7}{8} , \frac{8}{9} , \frac{9}{10}$$

Examples of Unsimplified Fractions

$$\frac{2}{4} , \frac{4}{6} , \frac{9}{12} , \frac{8}{10} , \frac{15}{18} , \frac{24}{28} , \frac{14}{16} , \frac{64}{72} , \frac{81}{90}$$

$$\frac{50}{60} = \frac{50 \div 10}{60 \div 10} = \frac{5}{6}$$

$$\frac{24}{30} = \frac{24 \div 2}{30 \div 2} = \frac{12}{15} = \frac{12 \div 3}{15 \div 3} = \frac{4}{5}$$

$$\frac{12}{60} = \frac{12 \div 2}{60 \div 2} = \frac{6}{30} = \frac{6 \div 6}{30 \div 6} = \frac{1}{5}$$

- A fraction is considered to be in simplified form when you cannot find a number that evenly divides into both the numerator and the denominator of the fraction.

  Note: The numerator is the top part of a fraction. The denominator is the bottom part of a fraction.

- A fraction is considered to be in unsimplified form when you can find a number that evenly divides into both the numerator and the denominator of the fraction.

1 To simplify, or reduce, a fraction, look for a number that evenly divides into both the numerator and the denominator of the fraction.

2 Divide the numerator and denominator of the fraction by the number you found in step 1.

3 Repeat steps 1 and 2 until you can no longer find numbers that will evenly divide into both the numerator and the denominator of the fraction.

Convert Improper Fractions to Mixed Numbers

Improper fractions aren't really that scandalous, they are just fractions with numerators, or top numbers, that are larger than the denominators, or bottom numbers, such as $\frac{3}{2}$, $\frac{64}{32}$, and $\frac{184}{5}$. Improper fractions are called improper because their value can be written in another way, which is easier to understand.

An improper fraction can be expressed as a mixed number—a whole number combined with a fraction, such as $1\frac{1}{2}$. Think of it this way. If you had to wait in line for half an hour to get into a concert, then at the concession stand for half an hour and then outside of the washroom for yet another half an hour, you wouldn't complain about being in line for three halves of an hour. You would say that you waited an hour and a half.

To convert an improper fraction to a mixed number, you need to divide the numerator of the improper fraction by the denominator. The answer becomes the whole number and the remainder becomes the new numerator. For example, in the improper fraction $\frac{7}{3}$, the number 3 divides into the number 7 two times and leaves a remainder of 1. The resulting mixed number would therefore be $2\frac{1}{3}$.

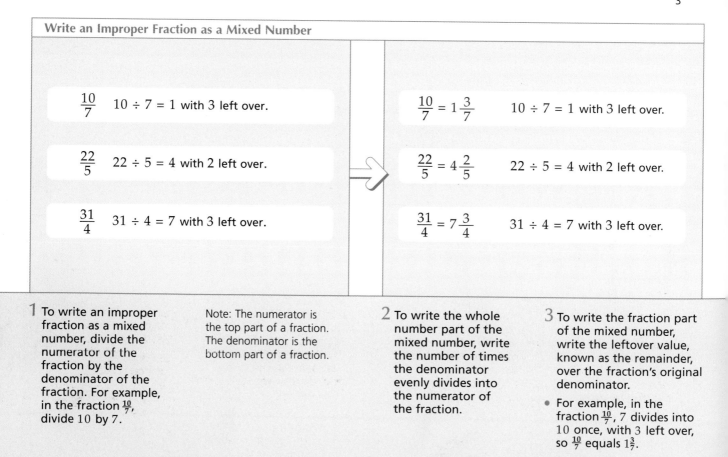

Write an Improper Fraction as a Mixed Number

$\frac{10}{7}$ $10 \div 7 = 1$ with 3 left over.

$\frac{22}{5}$ $22 \div 5 = 4$ with 2 left over.

$\frac{31}{4}$ $31 \div 4 = 7$ with 3 left over.

$\frac{10}{7} = 1\frac{3}{7}$ $10 \div 7 = 1$ with 3 left over.

$\frac{22}{5} = 4\frac{2}{5}$ $22 \div 5 = 4$ with 2 left over.

$\frac{31}{4} = 7\frac{3}{4}$ $31 \div 4 = 7$ with 3 left over.

1 To write an improper fraction as a mixed number, divide the numerator of the fraction by the denominator of the fraction. For example, in the fraction $\frac{10}{7}$, divide 10 by 7.

Note: The numerator is the top part of a fraction. The denominator is the bottom part of a fraction.

2 To write the whole number part of the mixed number, write the number of times the denominator evenly divides into the numerator of the fraction.

3 To write the fraction part of the mixed number, write the leftover value, known as the remainder, over the fraction's original denominator.

- For example, in the fraction $\frac{10}{7}$, 7 divides into 10 once, with 3 left over, so $\frac{10}{7}$ equals $1\frac{3}{7}$.

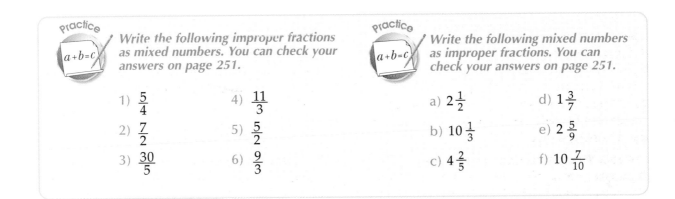

Practice

Write the following improper fractions as mixed numbers. You can check your answers on page 251.

1) $\frac{5}{4}$ 4) $\frac{11}{3}$

2) $\frac{7}{2}$ 5) $\frac{5}{2}$

3) $\frac{30}{5}$ 6) $\frac{9}{3}$

Practice

Write the following mixed numbers as improper fractions. You can check your answers on page 251.

a) $2\frac{1}{2}$ d) $1\frac{3}{7}$

b) $10\frac{1}{3}$ e) $2\frac{5}{9}$

c) $4\frac{2}{5}$ f) $10\frac{7}{10}$

Write a Mixed Number as an Improper Fraction

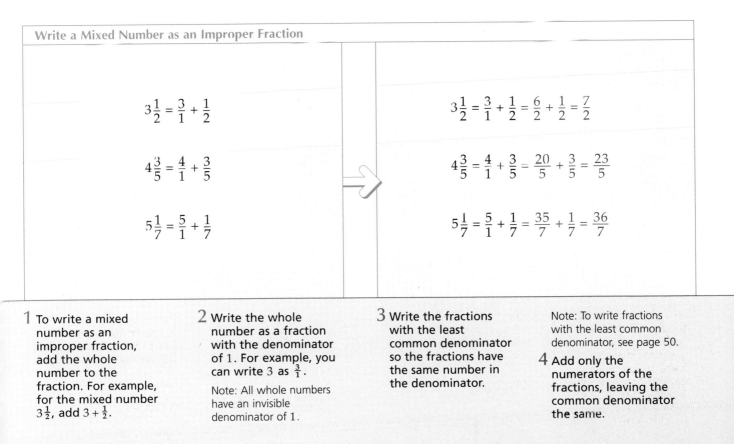

$$3\frac{1}{2} = \frac{3}{1} + \frac{1}{2}$$

$$4\frac{3}{5} = \frac{4}{1} + \frac{3}{5}$$

$$5\frac{1}{7} = \frac{5}{1} + \frac{1}{7}$$

$$3\frac{1}{2} = \frac{3}{1} + \frac{1}{2} = \frac{6}{2} + \frac{1}{2} = \frac{7}{2}$$

$$4\frac{3}{5} = \frac{4}{1} + \frac{3}{5} = \frac{20}{5} + \frac{3}{5} = \frac{23}{5}$$

$$5\frac{1}{7} = \frac{5}{1} + \frac{1}{7} = \frac{35}{7} + \frac{1}{7} = \frac{36}{7}$$

1 To write a mixed number as an improper fraction, add the whole number to the fraction. For example, for the mixed number $3\frac{1}{2}$, add $3 + \frac{1}{2}$.

2 Write the whole number as a fraction with the denominator of 1. For example, you can write 3 as $\frac{3}{1}$.

Note: All whole numbers have an invisible denominator of 1.

3 Write the fractions with the least common denominator so the fractions have the same number in the denominator.

Note: To write fractions with the least common denominator, see page 50.

4 Add only the numerators of the fractions, leaving the common denominator the same.

Multiply and Divide Fractions

Multiplying and dividing fractions are quite easy operations to master. In fact, you can put your division toolkit away for a while, since you are only going to need to multiply. That's right, due to a quirky characteristic of fractions, dividing one fraction by another fraction is the same as multiplying the first fraction by the reciprocal, or flipped version, of the second fraction. For example, $\frac{1}{2} \div \frac{3}{4}$ becomes $\frac{1}{2} \times \frac{4}{3}$ —no division skills needed!

You can also save a lot of time by working with more than two fractions at once. For example,

$$\frac{1}{2} \times \frac{3}{4} \times \frac{5}{6} = \frac{1 \times 3 \times 5}{2 \times 4 \times 6} \quad \text{and} \quad \frac{1}{2} \div \frac{3}{4} \div \frac{5}{6} = \frac{1 \times 4 \times 6}{2 \times 3 \times 5}$$

Remember that if you need to work with a fraction and a whole number, you can write a whole number as a fraction with its normally invisible denominator of 1. For example, you can write 2 as $\frac{2}{1}$.

Multiply Fractions

$$\frac{3}{4} \times \frac{6}{8} = \frac{3 \times 6}{4 \times 8} = \frac{18}{32}$$

$$\frac{5}{6} \times \frac{4}{7} = \frac{5 \times 4}{6 \times 7} = \frac{20}{42}$$

$$\frac{1}{2} \times \frac{2}{3} \times \frac{3}{4} = \frac{1 \times 2 \times 3}{2 \times 3 \times 4} = \frac{6}{24}$$

→

$$\frac{3}{4} \times \frac{6}{8} = \frac{3 \times 6}{4 \times 8} = \frac{18}{32} = \frac{9}{16}$$

$$\frac{5}{6} \times \frac{4}{7} = \frac{5 \times 4}{6 \times 7} = \frac{20}{42} = \frac{10}{21}$$

$$\frac{1}{2} \times \frac{2}{3} \times \frac{3}{4} = \frac{1 \times 2 \times 3}{2 \times 3 \times 4} = \frac{6}{24} = \frac{1}{4}$$

---

1 To multiply fractions, multiply the numerators of all the fractions and then multiply the denominators of all the fractions. The answer will be the numerator result over the denominator result.

Note: The numerator is the top number in a fraction. The denominator is the bottom number in a fraction.

2 Write the resulting fraction in simplified form.

Note: A fraction is considered to be in simplified form when you cannot find a number that evenly divides into both the numerator and the denominator. For more information on simplifying fractions, see page 45.

Can I use any shortcuts when working with large fractions?

If you write large fractions in simplified form before you multiply the fractions, the numbers in the fraction will be smaller and much easier to work with. For example, you can simplify $\frac{9}{36} \times \frac{12}{72}$ to $\frac{1}{4} \times \frac{1}{6}$.

$a + b = c$

Multiply or divide the following fractions and write each answer in simplified form. You can check your answers on page 251.

1) $\frac{1}{2} \times \frac{3}{4}$

2) $\frac{2}{3} \times 7$

3) $\frac{2}{5} \div 4$

4) $\frac{3}{5} \div \frac{2}{7}$

5) $\frac{1}{2} \times \frac{5}{6} \times \frac{3}{4}$

6) $\frac{1}{2} \times \frac{5}{6} \div \frac{3}{4}$

Divide Fractions

$$\frac{1}{2} \div \frac{3}{4} = \frac{1}{2} \times \frac{4}{3}$$

$$\frac{2}{3} \div \frac{6}{8} = \frac{2}{3} \times \frac{8}{6}$$

$$\frac{2}{5} \div \frac{5}{9} = \frac{2}{5} \times \frac{9}{5}$$

$$\frac{1}{2} \div \frac{3}{4} = \frac{1}{2} \times \frac{4}{3} = \frac{1 \times 4}{2 \times 3} = \frac{4}{6} = \frac{2}{3}$$

$$\frac{2}{3} \div \frac{6}{8} = \frac{2}{3} \times \frac{8}{6} = \frac{2 \times 8}{3 \times 6} = \frac{16}{18} = \frac{8}{9}$$

$$\frac{2}{5} \div \frac{5}{9} = \frac{2}{5} \times \frac{9}{5} = \frac{2 \times 9}{5 \times 5} = \frac{18}{25}$$

1 To divide fractions, you first need to flip the second fraction, placing the bottom number of the fraction on top and the top number on the bottom. This is known as finding the reciprocal of the fraction. For example, a fraction would change from $\frac{3}{4}$ to $\frac{4}{3}$.

2 Change the division sign ($\div$) to a multiplication sign ($\times$). You can now multiply the fractions.

3 To multiply fractions, multiply the numerators of all the fractions and then multiply the denominators of all the fractions. The answer will be the numerator result over the denominator result.

4 Write the resulting fraction in simplified form.

Note: A fraction is considered to be in simplified form when you cannot find a number that evenly divides into both the numerator and the denominator. For more information on simplifying fractions, see page 45.

Find the Least Common Denominator

Fractions, like people, work well together when they share something in common. In the case of fractions, you must find a common denominator—the bottom number of a fraction—in order to add, subtract or accurately compare a group of fractions.

Finding the Least Common Denominator, or LCD for short, is one of those basic algebra skills you will need to master. The idea of the least common denominator is to find the smallest denominator shared by a group of dissimilar fractions. Once you have found the least common denominator for a

group of fractions, you must write the fractions so they display a new numerator, or top number, to correspond to the new denominator. Keep in mind that fractions can be written different ways and still have the same value. For example, $\frac{3}{5}$ can be rewritten as $\frac{6}{10}$ or $\frac{27}{45}$ without changing the value of the fraction.

You can actually work with any denominator common to two or more fractions, but finding the least common denominator allows you to calculate with smaller numbers.

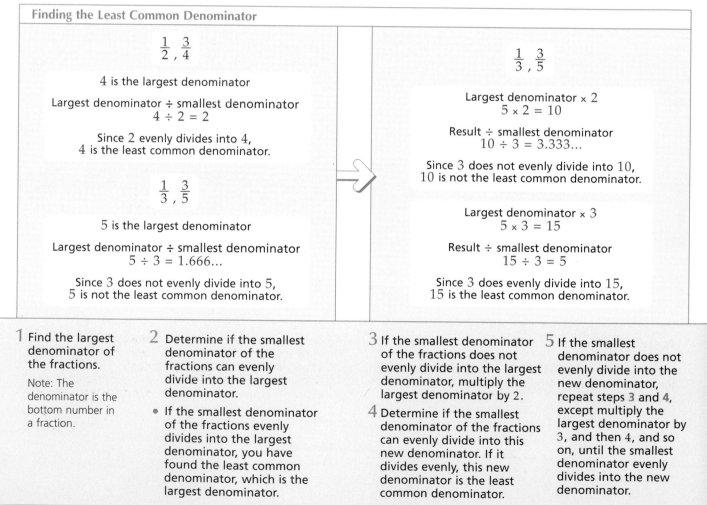

Finding the Least Common Denominator

$$\frac{1}{2} , \frac{3}{4}$$

4 is the largest denominator

Largest denominator ÷ smallest denominator
$$4 \div 2 = 2$$

Since 2 evenly divides into 4,
4 is the least common denominator.

$$\frac{1}{3} , \frac{3}{5}$$

5 is the largest denominator

Largest denominator ÷ smallest denominator
$$5 \div 3 = 1.666...$$

Since 3 does not evenly divide into 5,
5 is not the least common denominator.

$$\frac{1}{3} , \frac{3}{5}$$

Largest denominator × 2
$$5 \times 2 = 10$$

Result ÷ smallest denominator
$$10 \div 3 = 3.333...$$

Since 3 does not evenly divide into 10,
10 is not the least common denominator.

Largest denominator × 3
$$5 \times 3 = 15$$

Result ÷ smallest denominator
$$15 \div 3 = 5$$

Since 3 does evenly divide into 15,
15 is the least common denominator.

1 Find the largest denominator of the fractions.

Note: The denominator is the bottom number in a fraction.

2 Determine if the smallest denominator of the fractions can evenly divide into the largest denominator.

- If the smallest denominator of the fractions evenly divides into the largest denominator, you have found the least common denominator, which is the largest denominator.

3 If the smallest denominator of the fractions does not evenly divide into the largest denominator, multiply the largest denominator by 2.

4 Determine if the smallest denominator of the fractions can evenly divide into this new denominator. If it divides evenly, this new denominator is the least common denominator.

5 If the smallest denominator does not evenly divide into the new denominator, repeat steps 3 and 4, except multiply the largest denominator by 3, and then 4, and so on, until the smallest denominator evenly divides into the new denominator.

Tip

What is the quickest way to find a common denominator?

You can quickly find a common denominator by multiplying the denominators of two fractions together. The result may not be the least common denominator and you may end up working with large, awkward numbers, but this method provides speedy results. For example, for the fractions $\frac{5}{7}$ and $\frac{8}{9}$, you could multiply 7×9 to give you a common denominator of 63. If you use this method to find a common denominator, don't forget to simply the fractions in your final answer.

Practice

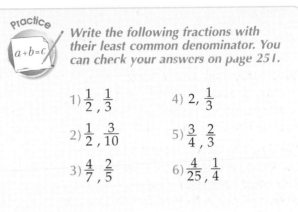

Write the following fractions with their least common denominator. You can check your answers on page 251.

1) $\frac{1}{2}$, $\frac{1}{3}$ 4) 2, $\frac{1}{3}$

2) $\frac{1}{2}$, $\frac{3}{10}$ 5) $\frac{3}{4}$, $\frac{2}{3}$

3) $\frac{4}{7}$, $\frac{2}{5}$ 6) $\frac{4}{25}$, $\frac{1}{4}$

Writing Fractions with the Least Common Denominator

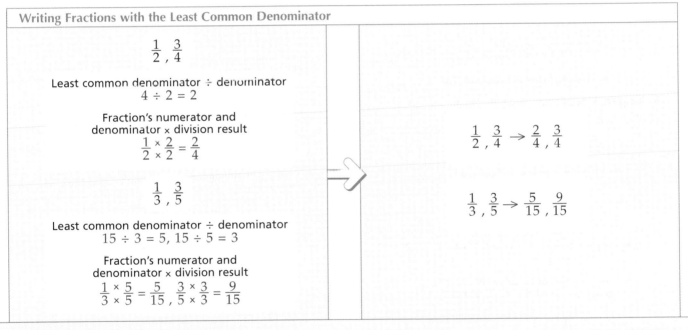

$$\frac{1}{2} , \frac{3}{4}$$

Least common denominator ÷ denominator
$$4 \div 2 = 2$$

Fraction's numerator and denominator × division result
$$\frac{1}{2} \times \frac{2}{2} = \frac{2}{4}$$

$$\frac{1}{3} , \frac{3}{5}$$

Least common denominator ÷ denominator
$$15 \div 3 = 5, \quad 15 \div 5 = 3$$

Fraction's numerator and denominator × division result
$$\frac{1}{3} \times \frac{5}{5} = \frac{5}{15}, \quad \frac{3}{5} \times \frac{3}{3} = \frac{9}{15}$$

$$\frac{1}{2} , \frac{3}{4} \rightarrow \frac{2}{4} , \frac{3}{4}$$

$$\frac{1}{3} , \frac{3}{5} \rightarrow \frac{5}{15} , \frac{9}{15}$$

1 To write fractions with the least common denominator, divide the least common denominator of the fractions by the denominator in each fraction.

Note: If a fraction's denominator is equal to the least common denominator, you do not need to perform step **1** since the fraction is already written with the least common denominator.

2 Multiply both the numerator and the denominator in each fraction by the number you determined in step **1**.

• You can now write the fractions with the least common denominator.

• After you write fractions with the least common denominator, you can easily add, subtract and compare the fractions.

Add and Subtract Fractions

Adding and subtracting fractions is not always straightforward. Most people can add $\frac{1}{2} + \frac{1}{2}$ in their heads, but adding $\frac{1}{2} + \frac{1}{3} + \frac{1}{4} + \frac{1}{5}$ is likely to induce a migraine. In order to add or subtract fractions, the fractions must all have a common denominator—in other words, the numbers on the bottom of the fractions must be the same. If your fractions do not conveniently come with common denominators, you will have to do a bit of work before you can add or subtract them.

To give each of your fractions a common denominator, you must multiply both the numerator, or top number, and denominator, or bottom number, of at least one of the fractions by the same number so that the denominators of the fractions will match. For instance, if you want to subtract $\frac{1}{4}$ from $\frac{1}{2}$, you will have to multiply both the numerator and denominator of $\frac{1}{2}$ by 2 to give you $\frac{2}{4}$. Then both fractions will have 4 as the common denominator.

Once you have a denominator that is common to all the fractions, you simply add or subtract the numerators, writing the result over the common denominator.

$$\frac{1}{2} + \frac{3}{5} = \frac{5}{10} + \frac{6}{10}$$

$$2 + \frac{2}{3} + \frac{4}{6} = \frac{2}{1} + \frac{2}{3} + \frac{4}{6}$$
$$= \frac{12}{6} + \frac{4}{6} + \frac{4}{6}$$

$$\frac{3}{4} - \frac{1}{2} = \frac{3}{4} - \frac{2}{4}$$

$\Rightarrow$

$$\frac{1}{2} + \frac{3}{5} = \frac{5}{10} + \frac{6}{10} = \frac{11}{10}$$

$$2 + \frac{2}{3} + \frac{4}{6} = \frac{2}{1} + \frac{2}{3} + \frac{4}{6}$$
$$= \frac{12}{6} + \frac{4}{6} + \frac{4}{6} = \frac{20}{6} = \frac{10}{3}$$

$$\frac{3}{4} - \frac{1}{2} = \frac{3}{4} - \frac{2}{4} = \frac{1}{4}$$

1 Write the fractions with their least common denominator, so that each fraction has the same denominator.

Note: The denominator is the bottom number in a fraction. To write fractions with their least common denominator, see page 50.

• When adding or subtracting a fraction with a whole number, you can write the number as a fraction. All whole numbers have an invisible denominator of 1. For example, you can write 2 as $\frac{2}{1}$.

2 Add or subtract only the numerators of the fractions, writing the result over the common denominator.

Note: The numerator is the top number in a fraction.

3 Make sure the resulting fraction is written in simplified form.

Note: A fraction is considered to be in simplified form when you cannot find a number that evenly divides into both the numerator and the denominator of the fraction. For more information on simplifying fractions, see page 45.

Write Numbers in Scientific Notation

It is estimated that there are 2 quadrillion (2,000,000,000,000,000) grains of sand on the world's beaches. As you can see, large numbers like 2 quadrillion take up a lot of space. Working with large numbers in long form would quickly become tiresome and use up a forest of paper in no time.

Fortunately for the trees, mathematicians use a kind of shorthand for writing very large or very small numbers, which is called scientific notation. Scientific notation uses exponents to express large and small numbers in a more compact form. Remember that exponents are those small numbers that appear above and to the right of a number

that indicate the number of times that number is multiplied by itself.

Scientific notation expresses numbers as a single, non-zero digit, often followed by a decimal place and multiplied by a power of 10. Using scientific notation, 2 quadrillion can be written as 2×10^{15}. Small numbers, such as .0000000034 are expressed as being multiplied by negative powers of ten, like 3.4×10^{-9}.

When writing a number in scientific notation, do not include any 0s at the end of the number being multiplied by the power of 10. For example, you should write 7.53×10^4 instead of 7.5300×10^4.

Write Large Numbers in Scientific Notation

$$70000 = 7 \times 10^4$$

$$600000 = 6 \times 10^5$$

$$2350000 = 2.35 \times 10^6$$

$$83900000 = 8.39 \times 10^7$$

$$325000 = 3.25 \times 10^5$$

Write Small Numbers in Scientific Notation

$$0.0009 = 9 \times 10^{-4}$$

$$0.00004 = 4 \times 10^{-5}$$

$$0.00000356 = 3.56 \times 10^{-6}$$

$$0.000051 = 5.1 \times 10^{-5}$$

$$0.0079 = 7.9 \times 10^{-3}$$

---

1 To write a large number in scientific notation, start at the end of the number and move the decimal place to the left, counting each number you pass, until only one number remains to the left of the decimal place.

2 Write the remaining number(s) followed by $\times 10^n$, where n is the number of numbers you passed.

- For example, for the number 70000, you would pass 4 numbers, so you would write this number in scientific notation as 7×10^4.

1 To write a small number in scientific notation, start at the decimal place and move the decimal place to the right, counting each number you pass, until one number that is greater than 0 is to the left of the decimal place.

2 Write the remaining number(s) followed by $\times 10^{-n}$, where n is the number of numbers you passed.

- For example, for the number 0.0009, you would pass 4 numbers, so you would write this number in scientific notation as 9×10^{-4}.

Exponent Basics

Exponents, those tiny numbers you see raised up on the right side of a number or variable, are clean-up specialists. Exponents are used to shorten expressions that contain repeated multiplications of the same number or variable and have tidied up many messy algebra expressions. Say you have 5 multiplied by itself four times. You could write it out as the bulky expression $5 \times 5 \times 5 \times 5$, but that is not a pretty sight. Through the miracle of exponents, however, the same expression becomes a pocket-sized 5^4.

The number or variable being multiplied by itself in an exponential expression is called the "base" of the exponent. When a number or variable is expressed with an exponent, the number or variable is said to be "raised to the power" of the exponent. For example, 2^4 is read as "two raised to the fourth power" or "two to the fourth power." Truly hip mathematicians will simply say "two to the fourth."

About Exponents

$2^4 = 2 \times 2 \times 2 \times 2 = 16$

$3^5 = 3 \times 3 \times 3 \times 3 \times 3 = 243$

$4^3 = 4 \times 4 \times 4 = 64$

$5^2 = 5 \times 5 = 25$

$6^2 = 6 \times 6 = 36$

Negative Numbers with Exponents

$(-2)^2 = (-2) \times (-2) = 4$

$(-2)^3 = (-2) \times (-2) \times (-2) = -8$

$(-2)^4 = (-2) \times (-2) \times (-2) \times (-2) = 16$

$(-2)^5 = (-2) \times (-2) \times (-2) \times (-2) \times (-2) = -32$

$(-2)^6 = (-2) \times (-2) \times (-2) \times (-2) \times (-2) \times (-2) = 64$

- An exponent, or power, appears as a small number above and to the right of a number, such as 2^4.

  Note: When an exponent cannot be shown as a small number above and to the right of a number, such as in an e-mail message, the exponent can be shown as 2^4.

- An exponent indicates the number of times a number is multiplied by itself. For example, 2^4 equals $2 \times 2 \times 2 \times 2$.

- The number to the left of an exponent is called the base.

- A negative number with an even-numbered exponent will result in a positive number. For example, $(-2)^2$ and $(-2)^4$ will both result in a positive number.

- A negative number with an odd-numbered exponent will result in a negative number. For example, $(-2)^3$ and $(-2)^5$ will both result in a negative number.

Tip

What do the terms "squared" and "cubed" mean?

When a number or variable is said to be "squared," it means the number or variable is raised to the second power. For instance, 7^2 is usually read as "7 squared." When a number or variable is said to be "cubed," it means the number or variable is raised to the third power. For instance, b^3 is usually read as "b cubed."

Practice

$a + b = c$

Evaluate the following problems that use exponents. You can check your answers on page 251.

1) 2^4

2) $3^2 5^3$

3) $(-3)^3$

4) -4^2

5) $(3 \times 7)^2$

6) 7^5

Exponents and Parentheses

$$-2^3 - -(2 \times 2 \times 2) = -8$$

$$(-2)^3 = (-2) \times (-2) \times (-2) = -8$$

$$(-3)^4 = (-3) \times (-3) \times (-3) \times (3) - 81$$

$$40 - 5^2 = 40 - (5 \times 5) = 40 - 25 = 15$$

$$40 + (-5)^2 = 40 + (-5) \times (-5) = 40 + (25) = 65$$

Variables with Exponents

$$x^4 = x \times x \times x \times x$$

$$x^2 = x \times x$$

$$(-x)^3 = (-x) \times (-x) \times (-x)$$

$$(-x)^4 = (-x) \times (-x) \times (-x) \times (-x)$$

$$-x^4 = -(x \times x \times x \times x)$$

- When a negative sign (–) appears in front of a number with an exponent, parentheses ( ) may appear around the number.

- If parentheses do not appear around a negative number with an exponent, ignore the negative sign (–) in front of the number when applying the exponent to the number. For example, -2^2 equals $-(2 \times 2)$.

- If parentheses appear around a negative number with an exponent, include the negative sign (–) in front of the number when applying the exponent to the number. For example, $(-2)^2$ equals $(-2) \times (-2)$.

- Variables can also have exponents.

- As with numbers, the exponent indicates the number of times a variable is multiplied by itself. For example, x^4 equals $x \times x \times x \times x$.

- The variable to the left of an exponent is called the base.

Rules for Exponents

You will find exponents lurking around many corners in algebra. If you don't know how to deal with tricky exponential expressions, things can quickly spiral into confusion.

Fortunately, the rules for working with exponents are not difficult to master. Working out the power of 1, for instance, is a simple task. Any number or variable raised to the power of one simply equals itself. So 12^1 can, and should, be simplified to 12. Including the exponent of 1 is not necessary.

The power of 0 is simple too. Any number or variable, except for 0, raised to the power of 0 equals 1. For instance, even a large base number like 8,131,972 raised to the power of 0 simply equals 1.

Simple rules also exist for multiplying and dividing exponents that have the same base numbers or variables. When multiplying exponents with the same base, simply add the exponents together. For example, $5^6 \times 5^3 = 5^{6+3} = 5^9$. Similarly, when dividing exponents with the same base, simply subtract the exponents. For example, $5^6 \div 5^3 = 5^{6-3} = 5^3$.

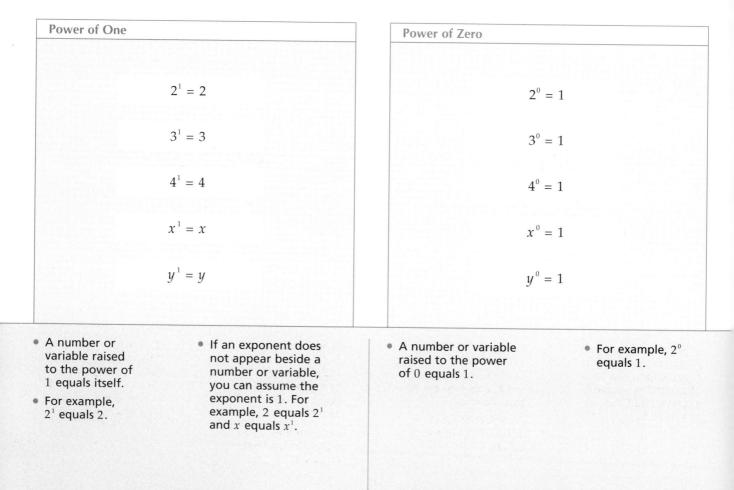

Power of One
$2^1 = 2$
$3^1 = 3$
$4^1 = 4$
$x^1 = x$
$y^1 = y$

Power of Zero
$2^0 = 1$
$3^0 = 1$
$4^0 = 1$
$x^0 = 1$
$y^0 = 1$

- A number or variable raised to the power of 1 equals itself.
- For example, 2^1 equals 2.

- If an exponent does not appear beside a number or variable, you can assume the exponent is 1. For example, 2 equals 2^1 and x equals x^1.

- A number or variable raised to the power of 0 equals 1.

- For example, 2^0 equals 1.

Tip

Is there a solution for 0^0?

No. While any other number or variable raised to the power of 0 equals 1, the expression 0^0 is said to be "undefined." This is how mathematicians say, "don't bother, it can't be done."

Practice

$a+b=c$

Simplify the following expressions using the rules for exponents. You can check your answers on page 252.

1) $4^2 \times 4^3$

2) $\dfrac{x^{11}}{x^4}$

3) $2^3 \times 2^2 \div 2^5$

4) $3^3 \times 4^3 \times 3^8 \times 4^5$

5) $y^2 \times y^7 \div y^4$

6) $10^4 \div 10^2 \times 10^{10}$

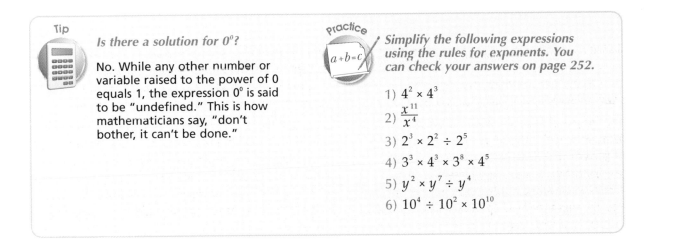

Multiplying Numbers with Exponents

$$2^3 \times 2^4 = 2^{3+4} = 2^7$$

$$3^4 \times 3^{-2} \times 3^5 = 3^{4+(-2)+5} = 3^7$$

$$5^{-6} \times 5^{-3} = 5^{(-6)+(-3)} = 5^{-9}$$

$$x^2 \times x^3 = x^{2+3} = x^5$$

$$y^5 \times y^{-2} \times y^3 = y^{5+(-2)+3} = y^6$$

Dividing Numbers with Exponents

$$\frac{2^8}{2^6} = 2^{8-6} = 2^2$$

$$\frac{4^7}{4^{-3}} = 4^{7-(-3)} = 4^{7+3} = 4^{10}$$

$$\frac{x^5}{x^1} = x^{5-1} = x^4$$

$$\frac{x^9}{x^{-4}} = x^{9-(-4)} = x^{9+4} = x^{13}$$

- When multiplying numbers or variables with exponents that have the same base, you can add the exponents.

Note: The number to the left of an exponent is called the base. For example, 2^3 and 2^4 both have the same base, which is 2.

- For example, $2^3 \times 2^4$ equals 2^{3+4}, which equals 2^7.

- When dividing numbers or variables with exponents that have the same base, you can subtract the exponents.

Note: When subtracting, if two negative (–) signs appear side by side, you can replace the signs with a single plus sign (+). For example, $7 -(-3)$ equals $7 + 3$.

- For example, $2^8 \div 2^6$ equals 2^{8-6}, which equals 2^2.

CONTINUED ▶

Rules for Exponents
continued

A number or variable can be raised to two exponents. This type of expression is written with the help of parentheses, such as $(8^2)^3$. While the expression looks complicated, simplifying it is just a matter of multiplying the two exponents together. For example, $(8^2)^3 = 8^{2 \times 3} = 8^6$.

Exponents can also be applied to fractions. Again, simplifying these expressions is not that tough. You just need to apply the exponent to both the numerator and the denominator. For example $\left(\frac{1}{2}\right)^3 = \frac{1^3}{2^3}$.

With all this talk of exponents, you may have started wondering about negative exponents. Negative exponents are actually a clever way of writing fractions and decimals. For instance, a^{-2} can be written as $\frac{1}{a^2}$. Keep in mind that some teachers prefer to see a fraction rather than a negative exponent in a final simplified answer.

Numbers Raised to Two Exponents

$$(2^3)^3 = 2^{3 \times 3} = 2^9$$

$$(3^5)^{-2} = 3^{(5) \times (-2)} = 3^{-10}$$

$$(5^{-4})^{-6} = 5^{(-4) \times (-6)} = 5^{24}$$

$$(x^2)^5 = x^{2 \times 5} = x^{10}$$

$$(y^{-3})^{-8} = y^{(-3) \times (-8)} = y^{24}$$

Multiplication or Division Results Raised to an Exponent

$$(2 \times 3)^2 = 2^2 \times 3^2$$

$$\left(\frac{6}{2}\right)^2 = \frac{6^2}{2^2}$$

$$(3x)^4 = 3^4 \times x^4 = 81 \times x^4 = 81x^4$$

$$(xy)^3 = x^3 \times y^3 = x^3 y^3$$

$$\left(\frac{x}{y}\right)^3 = \frac{x^3}{y^3}$$

- When a number or variable is raised to two exponents, you can multiply the exponents together.

  Note: When a number or variable has an exponent, it is also said to be "raised to the power" of that exponent.

- For example, $(2^3)^3$ equals $2^{3 \times 3}$, which equals 2^9.

Note: When multiplying two numbers with a different sign (+ and −), the result will be a negative number. For example $4 \times (-6)$ equals −24. When multiplying two numbers with the same sign (+ or −), the result will be a positive number. For example $(-4) \times (-6)$ equals 24.

- When multiplying or dividing numbers or variables and the result is raised to an exponent, you can raise each number or variable to the same exponent.

- For example, $(2 \times 3)^2$ equals $2^2 \times 3^2$.

Note: If a symbol does not appear between a number and variable, such as $3x$, or two variables, such as xy, you multiply the number and variable or variables together. For example, $3x$ equals $3 \times x$ and xy equals $x \times y$.

Tip

How do I find the reciprocal of a number or variable?

Finding the reciprocal of a fraction is easy—just flip it over, switching the top and bottom numbers. For instance, the reciprocal of $\frac{2}{6}$ would be $\frac{6}{2}$. To find the reciprocal of a whole number or variable, write it as a fraction and then flip the fraction. Since all whole numbers and variables have an invisible denominator of 1, x can be written $\frac{x}{1}$, so the reciprocal would be $\frac{1}{x}$.

Practice

$a + b = c$

Simplify the following expressions using the rules for exponents. You can check your answers on page 252.

1) $(2^3)^5$

2) $(3^{-2})^5$

3) $(3 \times 5)^2$

4) $\left(\dfrac{7}{4}\right)^3$

5) $\left(\dfrac{9}{5}\right)^{-1}$

6) $\left(\dfrac{3}{4}\right)^{-2}$

Negative Exponents

$$2^{-4} = \frac{1}{2^4}$$

$$x^{-2} = \frac{1}{x^2}$$

$$\frac{1}{2^{-6}} = \frac{2^6}{1} = 2^6$$

$$\frac{1}{x^{-7}} = \frac{x^7}{1} = x^7$$

$$\frac{x^2}{y^4 z^{-3}} = \frac{x^2}{y^4} \times \frac{1}{z^{-3}} = \frac{x^2}{y^4} \times \frac{z^3}{1} = \frac{x^2 z^3}{y^4}$$

$$\frac{a^{-5} b^2}{c^4} = \frac{a^{-5}}{1} \times \frac{b^2}{c^4} = \frac{1}{a^5} \times \frac{b^2}{c^4} = \frac{b^2}{a^5 c^4}$$

$$\frac{4a^{-2}}{b^{-6}} = 4 \times \frac{a^{-2}}{1} \times \frac{1}{b^{-6}} = 4 \times \frac{1}{a^2} \times \frac{b^6}{1} = \frac{4b^6}{a^2}$$

$$(x - 5)^{-3} = \frac{1}{(x - 5)^3}$$

- If a number or variable has a negative exponent, such as 2^{-4} or x^{-2}, you can change the exponent to a positive exponent.

1 To change a negative exponent to a positive exponent, write the reciprocal of the number or variable before the exponent.

Note: To determine the reciprocal of a number or variable, see the top of this page.

2 Change the exponent from a negative number (–) to a positive number (+).

- When negative exponents appear in division problems, you can change the negative exponents to positive exponents using the steps you just learned.

Working With Fractions and Exponents

Question 1. Simplify the following fractions as much as possible.

a) $\dfrac{6}{3}$

b) $\dfrac{20}{30}$

c) $-\dfrac{3}{15}$

d) $\dfrac{400}{500}$

e) $-\dfrac{40}{100}$

f) $\dfrac{14}{21}$

Question 2. Convert the following improper fractions to mixed numbers.

a) $\dfrac{3}{2}$

b) $-\dfrac{10}{5}$

c) $-\dfrac{20}{3}$

d) $\dfrac{5}{2}$

e) $\dfrac{7}{5}$

Question 3. Convert the following mixed numbers to improper fractions.

a) $3\dfrac{1}{2}$

b) $2\dfrac{1}{3}$

c) $-2\dfrac{1}{9}$

d) $4\dfrac{1}{4}$

e) $5\dfrac{1}{10}$

Question 4. Perform the following operations.

a) $\dfrac{1}{2} \times \dfrac{1}{3}$

b) $\dfrac{2}{5} \times \dfrac{3}{7}$

c) $-\dfrac{1}{8} \times \dfrac{2}{9}$

d) $\dfrac{1}{4} \div \dfrac{1}{3}$

e) $\dfrac{3}{8} \div \left(-\dfrac{6}{7}\right)$

f) $\dfrac{1}{5} \div 4$

Question 5. Write the following sets of fractions with their least common denominator.

a) $\dfrac{1}{2}$, $\dfrac{1}{3}$

b) $\dfrac{1}{5}$, $\dfrac{1}{4}$

c) $\dfrac{3}{4}$, $-\dfrac{5}{6}$

d) $-\dfrac{1}{10}$, $\dfrac{3}{25}$

e) $-\dfrac{13}{100}$, $\dfrac{2}{5}$

Question 6. Perform the following operations.

a) $\dfrac{1}{2} + \dfrac{1}{5}$

b) $\dfrac{3}{4} + \dfrac{1}{2}$

c) $\dfrac{2}{5} - 1$

d) $-\dfrac{3}{7} + \dfrac{2}{3}$

e) $2 - \dfrac{1}{5}$

f) $-\dfrac{3}{8} - \dfrac{3}{4}$

Question 7. Write the following numbers in scientific notation.

a) 1,000

b) 0.001

c) 239,861

d) 1,023

e) 0.0000230

f) 1,000,130

g) 0.3

Question 8. Simplify the following expressions as much as possible using exponents.

a) $5^2 \times 5^4$

b) $3^4 \times 3^{-7}$

c) $\dfrac{2^8 \times 3^{-2} \times 2^{-2}}{3^{-5}}$

d) $\dfrac{10 \times 10 \times 10}{10}$

e) $\dfrac{6^4}{6^{-10}}$

f) $(xy)^2$

g) $(2 \times 5)^3$

You can check your answers on pages 267-268.

Equations are an integral part of algebra. Since you will be coming across equations at every turn, understanding how to work with them is essential. Chapter 3 provides you with everything you need to know for solving basic equations.

Solving Basic Equations

Solve an Equation with Addition and Subtraction

Addition and subtraction are the basic tools you will begin using to solve equations. Before you start adding and subtracting, however, you must understand the principle of keeping the equation balanced. Whenever you perform an operation on one side of the equals sign, you must perform the same operation on the other side so the equation will remain balanced and true.

The object of your adding and subtracting is to remove or "cancel out" numbers. Canceling out numbers isolates the variable to one side of the equation, usually the left side, and is referred to as "solving for the variable." For example, in the equation $x + 7 = 9$ you would subtract 7 from each side of the equals sign to cancel out the 7 originally found on the left side. In this example, we would have $x + 7 - 7 = 9 - 7$, which leaves an answer of $x = 2$.

Thankfully, equations are easy to check. All you have to do is take the value you determined for the variable, replace the variable in the original equation with the value you found and then work out the problem. If one side equals the other, you have correctly solved for the variable.

How to Solve an Equation

$$x - 5 = 11$$
$$x - 5 + 5 = 11 + 5$$
$$x = 16$$

Check your answer
$$x - 5 = 11$$
$$16 - 5 = 11$$
$$11 = 11 \quad \textit{Correct!}$$

Solving Equations Example 1

$$x + 7 = 20$$
$$x + 7 - 7 = 20 - 7$$
$$x = 13$$

Check your answer
$$x + 7 = 20$$
$$13 + 7 = 20$$
$$20 = 20 \quad \textit{Correct!}$$

- When solving equations, your goal is to place the variable on one side of the equation, usually the left side, and place all the numbers on the other side of the equation, usually the right side. This allows you to determine the value of the variable.

Note: A variable is a letter, such as x or y, which represents any number.

- To place a variable by itself on one side of an equation, you can add or subtract the same number or variable on both sides of the equation and the equation will remain balanced.

1 In this example, to place the variable by itself on the left side of the equation, subtract 7 from both sides of the equation. This will remove, or cancel out, 7 from the left side of the equation.

2 To check your answer, place the number you found into the original equation and solve the problem. If both sides of the equation are equal, you have correctly solved the equation.

Tip

How do I know when I should add or subtract to solve an equation?

As a simple rule of thumb, you should do the opposite of what is done in the original equation. When a number is subtracted from a variable in an equation, such as $x - 5 = 11$, you add the subtracted number to both sides of the equation to cancel out the number and isolate the variable. If a number is added to a variable, such as $x + 5 = 11$, you can subtract the added number from both sides of the equation to cancel out the number and isolate the variable.

Practice

$a + b = c$

Solve for the variable in each of the following equations. You can check your answers on page 252.

1) $x + 2 = 4$

2) $x - 3 = 10$

3) $5 = x - 4$

4) $-3 = x - 3$

5) $2x - 3 = x + 7$

6) $5x + 1 = 6x - 4$

Solving Equations Example 2

$$5 = 20 + x$$
$$5 - 20 = 20 + x - 20$$
$$-15 = x$$
$$x = -15$$

Check your answer
$$5 = 20 + x$$
$$5 = 20 + (-15)$$
$$5 = 5 \quad \text{Correct!}$$

1 In this example, to place the variable by itself on the right side of the equation, subtract 20 from both sides of the equation.

2 To check your answer, place the number you found into the original equation and solve the problem. If both sides of the equation are equal, you have correctly solved the equation.

Solving Equations Example 3

$$8x - 6 = 7x + 3$$
$$8x - 6 + 6 = 7x + 3 + 6$$
$$8x = 7x + 9$$
$$8x - 7x = 7x + 9 - 7x$$
$$x = 9$$

Check your answer
$$8x - 6 = 7x + 3$$
$$8(9) - 6 = 7(9) + 3$$
$$72 - 6 = 63 + 3$$
$$66 = 66 \quad \text{Correct!}$$

1 In this example, to place all the numbers on the right side of the equation, add 6 to both sides of the equation. You can then subtract $7x$ from both sides of the equation to move all the variables to the left side of the equation.

2 To check your answer, place the number you found into the original equation and solve the problem. If both sides of the equation are equal, you have correctly solved the equation.

Solve an Equation with Multiplication and Division

When multiplying and dividing in equations, you must remember the principle of balance, just like when you add and subtract within equations. If you perform an operation on one side of the equals sign, you have to perform the same operation on the other side in order for the equation to remain balanced and true.

Your aim in multiplying and dividing in an equation is to remove or "cancel out" numbers and isolate the variable to one side of the equation, usually the left side. This is called "solving for the variable."

When working with equations, you will often have to remove a coefficient, which is a number next to a variable, such as $4a$. Since the variable is being multiplied by the coefficient, you should divide both sides by the coefficient to cancel out the coefficient and isolate the variable.

To check your answers, take the value you determined for the variable, substitute it into the original equation and then work out the problem. If one side equals the other, you have correctly solved for the variable.

How to Solve an Equation

$$5x = 20$$
$$\frac{5x}{5} = \frac{20}{5}$$
$$x = 4$$

Check your answer
$$5x = 20$$
$$5(4) = 20$$
$$20 = 20 \quad \textit{Correct!}$$

Solving Equations Example 1

$$8x = 24$$
$$\frac{8x}{8} = \frac{24}{8}$$
$$x = 3$$

Check your answer
$$8x = 24$$
$$8(3) = 24$$
$$24 = 24 \quad \textit{Correct!}$$

- When solving an equation, your goal is to place the variable by itself on one side of the equation, usually the left side, and place all the numbers on the other side of the equation, usually the right side. This allows you to determine the value of the variable.

- To place a variable by itself on one side of an equation, you can multiply or divide by the same number on both sides of the equation and the equation will remain balanced.

1 In this example, to place the variable by itself on the left side of the equation, divide both sides of the equation by the number in front of the variable, called the coefficient. This will leave the x variable by itself on the left side of the equation because $\frac{8}{8}$ equals 1.

2 To check your answer, place the number you found into the original equation and solve the problem. If both sides of the equation are equal, you have correctly solved the equation.

Tip

How do I know when I should multiply or divide to solve an equation?

Generally, you should do the opposite of what is done in the original equation. When a variable is divided by a number in an equation, such as $\frac{x}{5} = 10$, you can multiply by the same number on both sides of the equation to cancel out the number and isolate the variable. If a variable is multiplied by a number, such as $5x = 20$, you can divide both sides of the equation by the same number to isolate the variable.

Practice

Solve for the variable in each of the following equations. You can check your answers on page 252.

1) $\frac{x}{2} = 1$

2) $4x = 16$

3) $-3x = 9$

4) $\frac{x}{2} = -5$

5) $\frac{x}{7} = 3$

6) $\frac{2x}{3} = \frac{5}{4}$

Solving Equations Example 2

$$\frac{x}{2} = 30$$

$$\frac{x}{2} \times 2 = 30 \times 2$$

$$x = 60$$

Check your answer

$$\frac{x}{2} = 30$$

$$\frac{60}{2} = 30$$

$$30 = 30 \quad \text{Correct!}$$

Solving Equations Example 3

$$\frac{2}{5}x = 6$$

$$\frac{2}{5}x \times \frac{5}{2} = 6 \times \frac{5}{2}$$

$$x = \frac{30}{2} = 15$$

Check your answer

$$\frac{2}{5}x = 6$$

$$\frac{2}{5}(15) = 6$$

$$\frac{30}{5} = 6$$

$$6 = 6 \quad \text{Correct!}$$

---

1 In this example, to place the variable by itself on the left side of the equation, you must multiply both sides of the equation by the number divided into the variable.

2 To check your answer, place the number you found into the original equation and solve the problem. If both sides of the equation are equal, you have correctly solved the equation.

1 In this example, to place the variable by itself on the left side of the equation, multiply both sides of the equation by the reciprocal of the fraction in front of the variable.

Note: To determine the reciprocal of a fraction, switch the top and bottom numbers in the fraction. For example, $\frac{2}{5}$ becomes $\frac{5}{2}$. When you multiply a fraction by its reciprocal, the result is 1.

2 To check your answer, place the number you found into the original equation and solve the problem. If both sides of the equation are equal, you have correctly solved the equation.

Solve an Equation in Several Steps

Even when you are working with large, complex problems, your goal should be to isolate the variable to one side of the equation and find the value of the variable. Three easy steps, using skills you have learned in the previous pages, will guide you when you are confronted with a complicated equation.

The first step in solving complex equations is to simplify the equation as much as you can on either side of the equals sign. You will find the distributive property particularly useful, as it eliminates those pesky parentheses. The second step involves using addition and subtraction on both sides of the equation to isolate the variable on one side of the equation. For the third step, you multiply and divide on both sides of the equation to determine the value of the variable.

Finally, don't forget the unofficial fourth step: check your answer! Place the number you determined for the variable into the original equation and solve the problem. If both sides of the equation are equal, you have correctly solved the equation.

Step 1: Simplify Each Side of the Equation

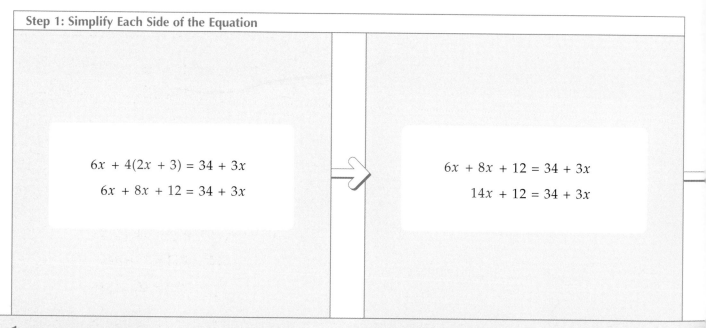

$$6x + 4(2x + 3) = 34 + 3x$$
$$6x + 8x + 12 = 34 + 3x$$

$$6x + 8x + 12 = 34 + 3x$$
$$14x + 12 = 34 + 3x$$

1 If an equation contains parentheses ( ), use the distributive property on each side of the equals sign (=) to remove the parentheses that surround numbers and variables.

- In this example, $4(2x + 3)$ works out to $8x + 12$.

Note: The distributive property allows you to remove a set of parentheses by multiplying each number or variable inside the parentheses by a number or variable directly outside the parentheses. For more information on the distributive property, see page 30.

2 Add, subtract, multiply and divide any numbers or variables that can be combined on the same side of the equals sign (=).

- In this example, $6x + 8x$ equals $14x$.

Note: For information on adding and subtracting variables, see page 36. For information on multiplying and dividing variables, see page 38.

Tip

What should I do if I end up with a negative variable?

If you end up with a negative sign (–) in front of a variable, such as –y = 14, you will have to make the variable positive to finish solving the equation. Fortunately, this is a simple task—just multiply both sides of the equation by –1. The equation remains balanced and a negative sign (–) will no longer appear in front of the variable. In the case of –y = 14, you would come out with a solution of y = –14.

Practice

Solve for the variable in each of the following equations. You can check your answers on page 252.

1) $3x = 6$

2) $4x - 2 = 14$

3) $2x + 10 = 3x - 5$

4) $x + 5 - 2x + 3 = 7 - 3x - 1$

5) $2(x - 1) = 3(x + 1) - 3$

6) $5(2x - 3) + x = 2(2 - x) + 7$

Step 2: Add and Subtract

$$14x + 12 = 34 + 3x$$

$$14x + 12 - 12 = 34 + 3x - 12$$

$$14x = 22 + 3x$$

$$14x - 3x = 22 + 3x - 3x$$

$$11x = 22$$

Step 3: Multiply and Divide

$$11x = 22$$

$$\frac{11x}{11} = \frac{22}{11}$$

$$x = 2$$

Check your answer

$$6x + 4(2x + 3) = 34 + 3x$$

$$6(2) + 4(2 \times 2 + 3) = 34 + 3(2)$$

$$12 + 4(7) = 34 + 6$$

$$40 = 40 \quad Correct!$$

---

3 Determine which numbers and variables you need to add or subtract to place the variable by itself on one side of the equation, usually the left side. Then add or subtract the same numbers and variables on both sides of the equation.

• In this example, subtract 12 from both sides of the equation to move all the numbers to the right side of the equation. You can then subtract $3x$ from both sides of the equation to place all the variables on the left side of the equation.

Note: For more information on adding and subtracting numbers and variables in equations, see page 64.

4 Determine which numbers you need to multiply or divide by to place the variable by itself on one side of the equation. Then multiply or divide by the same numbers on both sides of the equation.

• In this example, divide both sides of the equation by 11 to place the variable by itself on the left side of the equation. The variable x remains on the left side of the equation by itself because 11 ÷ 11 equals 1.

Note: For more information on multiplying and dividing numbers in equations, see page 66.

Solve for a Variable in an Equation with Multiple Variables

In algebra, it is common to come across equations that have more than one variable. An equation containing multiple variables, such as $a - 2 = 4 + b$, shows you the relationship between the variables. You can easily isolate any of the variables to one side of the equation to simplify the relationship. For example, if you know that $a = 6 + b$, you do not need to know the numerical value of a in order to define b. By simply rearranging the equation, you can find that $b = a - 6$.

Working with equations containing multiple variables is really no more difficult than working with equations that have just one variable. The only difference is that your final result will have variables on both sides of the equation.

Step 1: Simplify Each Side of the Equation

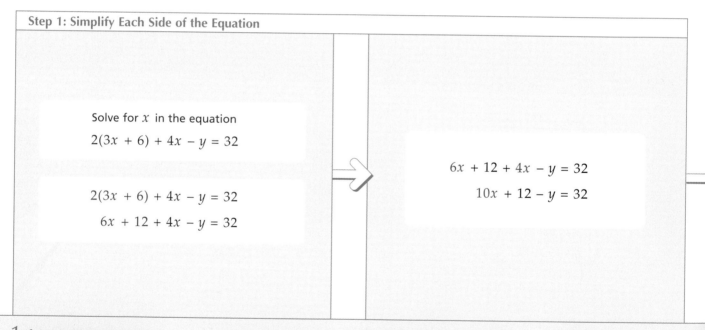

Solve for x in the equation

$$2(3x + 6) + 4x - y = 32$$

$$2(3x + 6) + 4x - y = 32$$
$$6x + 12 + 4x - y = 32$$

$$6x + 12 + 4x - y = 32$$
$$10x + 12 - y = 32$$

1 If an equation contains parentheses ( ), use the distributive property on each side of the equals sign (=) to remove the parentheses that surround numbers and variables.

- In this example, $2(3x + 6)$ works out to $6x + 12$.

Note: The distributive property allows you to remove a set of parentheses by multiplying each number or variable inside the parentheses by a number or variable directly outside the parentheses. For more information on the distributive property, see page 30.

2 Add, subtract, multiply and divide any numbers or variables that can be combined on each side of the equation.

- In this example, $6x + 4x$ equals $10x$.

Note: For information on adding and subtracting variables, see page 36. For information on multiplying and dividing variables, see page 38.

Tip

Once I have solved for a variable in an equation, do I need to further simplify the answer?

If you end up with a fraction that has two or more numbers and variables added or subtracted in the numerator, you can simplify your answer by splitting the fraction into smaller fractions. Each new fraction will contain one term of the original numerator and a duplicate of the denominator.

$$x = \frac{20 + y}{10} = \frac{20}{10} + \frac{y}{10} = 2 + \frac{y}{10}$$

Practice

Solve for x in the following equations. You can check your answers on page 252.

1. $2x + 3 - y = 5$
2. $x + 6y = 4 - x$
3. $3(y + 4) = 5 + x$

Solve for y in the following equations. You can check your answers on page 252.

4. $2(x - 1) + 2y = 5 + y$
5. $3(x + y) - 4y = 2$
6. $y + x + 2 = 4 - y + x$

Step 2: Add and Subtract

$$10x + 12 - y = 32$$
$$10x + 12 - y - 12 = 32 - 12$$
$$10x - y - 20$$
$$10x - y + y = 20 + y$$
$$10x = 20 + y$$

Step 3: Multiply and Divide

$$10x = 20 + y$$
$$\frac{10x}{10} = \frac{20 + y}{10}$$
$$x = \frac{20 + y}{10}$$

3 Determine which numbers and variables you need to add or subtract to place the variable by itself on one side of the equation, usually the left side.

4 Add or subtract the numbers and/or variables you determined in step 3 on both sides of the equation.

- In this example, subtract 12 from both sides of the equation. You can then add y to both sides of the equation.

Note: For information on adding and subtracting numbers and variables in equations, see page 64.

5 Determine which numbers you need to multiply or divide by to place the variable by itself on one side of the equation.

6 Multiply or divide by the numbers you determined in step 5 on both sides of the equation.

- In this example, divide both sides of the equation by 10. The x variable remains by itself on the left side of the equation because $10 \div 10$ equals 1.

Note: For information on multiplying and dividing numbers in equations, see page 66.

Solve an Equation with Absolute Values

Every once in a while you will come across an equation that contains an innocent-looking pair of absolute value symbols (| |) such as $x = |y + 3|$. These equations are known as absolute value equations and they are not as simple as they look.

Remember that the absolute value of a number or variable is always the positive value of that number or variable, whether the number or variable is positive or negative. So $|a|$ can equal either a or $-a$.

You will have to approach absolute value equations slightly differently than you would approach a normal equation because an absolute value equation may result in more than one answer. For example, the absolute value equation $|x| = 1 + 3$ can be simplified as $|x| = 4$. You can then determine that $x = 4$ or -4. You get two answers for the price of one!

Step 1: Isolate the Absolute Value Expression

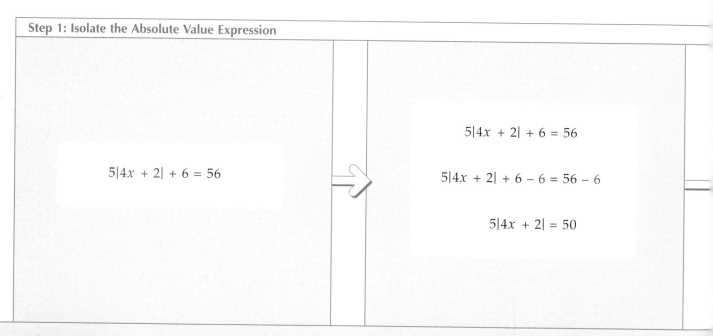

$$5|4x + 2| + 6 = 56$$

$$5|4x + 2| + 6 = 56$$

$$5|4x + 2| + 6 - 6 = 56 - 6$$

$$5|4x + 2| = 50$$

- When solving an equation that contains an absolute value expression, you first need to place the absolute value expression by itself on one side of the equation.

  Note: An absolute value expression consists of variables and/or numbers inside absolute value symbols ||. The result of an absolute value expression is always greater than zero or equal to zero. For more information on absolute values, see page 24.

1 Determine which numbers and variables you need to add or subtract to place the absolute value expression by itself on one side of the equation.

2 Add or subtract the numbers and variables you determined in step **1** on both sides of the equation.

- In this example, subtract 6 from both sides of the equation to remove 6 from the left side of the equation.

  Note: For information on adding and subtracting numbers and variables in equations, see page 64.

72

Tip

How do I solve a simple equation with absolute values?

Simple equations containing absolute values, like $|x| = 8$, are fairly simple to evaluate. Remember that all you are being asked for is the possible solutions for x. In the example $|x| = 8$, both 8 and −8 have an absolute value of 8, so $x = 8$ or −8.

Tip

Why do I need to create two equations from each equation that contains absolute values?

The key to solving an equation that contains absolute values is removing the absolute value symbols (||). Values within these symbols may have a negative or positive value, so you have to work out both possibilities by creating and solving two equations.

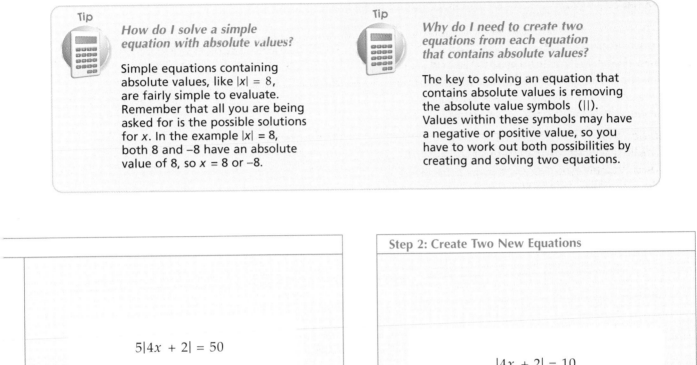

Step 2: Create Two New Equations

$$5|4x + 2| = 50$$

$$\frac{5|4x + 2|}{5} = \frac{50}{5}$$

$$|4x + 2| = 10$$

$$|4x + 2| = 10$$

$$4x + 2 = 10 \qquad 4x + 2 = -10$$

3 Determine which numbers you need to multiply or divide by to place the absolute value expression by itself on one side of the equation.

4 Multiply or divide by the numbers you determined in step 3 on both sides of the equation.

• In this example, divide both sides of the equation by 5 to remove 5 from the left side of the equation.

Note: For information on multiplying and dividing numbers in equations, see page 66.

• You now need to create two new equations from the original equation.

5 Create two new equations that look like the original equation, but without the absolute value symbols ||.

6 For the second equation, place a negative sign (−) in front of the value on the opposite side of the absolute value expression.

CONTINUED

Solve an Equation with Absolute Values *continued*

After you have created two new equations from the original absolute value equation, you can solve each of the two new equations and determine the two possible answers for the variable in an absolute value equation. The fun doesn't end there, though.

Even though it is twice the work, it is vital to check both of the numbers you found for the variable. When you place a number you found for the variable into the original equation and work out the equation, both sides of the equation must equal each other in order to have a solution.

Sometimes both of the numbers you found for the variable will solve the equation. In other instances, one or both of your numbers will fail to solve the equation. This means that either you have made a mistake and should double-check your calculations or there is only one or no solution for the equation.

Step 3: Solve the New Equations

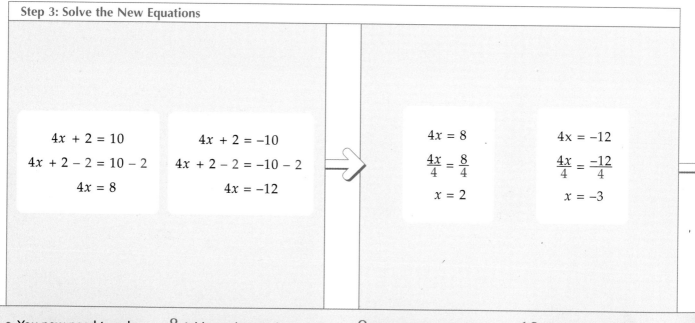

- You now need to solve the two new equations separately in order to determine the answer for the original equation.

7 In both equations, determine which numbers and variables you need to add or subtract to place the variable by itself on one side of the equation.

8 Add or subtract the numbers and/or variables you determined in step 7 on both sides of the equation.

- In both examples, subtract 2 from both sides of the equation.

9 In both equations, determine which numbers you need to multiply or divide by to place the variable by itself on one side of the equation.

10 Multiply or divide by the numbers you determined in step 9 on both sides of the equation.

- In both examples, divide by 4 on both sides of the equation.

Tip

Is there a solution to every equation with absolute values?

No. As you work with equations that contain absolute values, you will sometimes come across equations that have no solution. One dead give-away is when the absolute value equals a negative number, as in $|4x + 2| = -10$. Don't even bother working out these equations because no matter what x equals, the absolute value of the expression within the absolute value symbols cannot equal a negative number.

Practice

$a + b = c$

Solve the following equations containing absolute values. You can check your answers on page 252.

1) $|x| = 5$

2) $|2x - 3| = 5$

3) $2|x - 1| = 8$

4) $2|3x - 4| - 3 = 5$

5) $|3x - 9| = 0$

6) $3|2x + 6| + 5 = 2$

Step 4: Check Your Answers

If $x = 2$

$5|4x + 2| + 6 = 56$

$5|4(2) + 2| + 6 = 56$

$5|8 + 2| + 6 = 56$

$5|10| + 6 = 56$

$5 \times 10 + 6 = 56$

$56 = 56$ *Correct!*

If $x = -3$

$5|4x + 2| + 6 = 56$

$5|4(-3) + 2| + 6 = 56$

$5|-12 + 2| + 6 = 56$

$5|-10| + 6 = 56$

$5 \times 10 + 6 = 56$

$56 = 56$ *Correct!*

11 To check your answers, place the first number you found into the original equation and solve the problem. If both sides of the equation are equal, this number correctly solves the equation.

12 Place the second number you found into the original equation and solve the problem. If both sides of the equation are equal, this number also correctly solves the equation.

Solving Basic Equations

Question 1. Solve the following equations.

a) $2 + x = 5$

b) $x - 4 = 10$

c) $x + 3 = 0$

d) $2x = -8$

e) $3x = 5$

f) $-x = -2$

g) $-8x = -5$

h) $\dfrac{x}{4} = 6$

i) $\dfrac{5x}{3} = 10$

j) $2x = 1$

k) $\dfrac{3x}{7} = 3$

Question 2. Solve the following equations.

a) $2x + x + 1 = 2$

b) $2x + 1 = x$

c) $2x + 3 = 4x + 4$

d) $20 - x(4 + 1) = 5x$

e) $8x - 2(5x + 6) = 20$

f) $5x + 4(2x - 7) = -15$

Question 3. Solve for y in the following equations.

a) $y + x = 0$

b) $x + y = 2 + x$

c) $2y + 3x = 5 - x$

d) $x - 1 = y + 4$

e) $y + x + y + 1 = 2x - 3y + 8$

f) $2(x - y) = 3(x + y + 1)$

Question 4. Solve the following equations.

    a) $|x| = 3$

    b) $|-x| = 2$

    c) $|x| = 0$

    d) $|x| = -4$

    e) $|x - 2| = 1$

    f) $|x - 10| = 3$

    g) $|x + 5| = 5$

Question 5. Solve the following equations.

    a) $|2x| = 4$

    b) $|3x| = 8$

    c) $|2x - 1| = -1$

    d) $|3x - 6| = 0$

    e) $|2x - 1| = 1$

    f) $|5x - 2| + 4 = 6$

    g) $2|3x - 1| = 20$

    h) $4|2x + 3| + 4 = 20$

*You can check your answers
on pages 268-269.*

Chapter 4

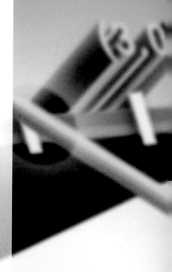

Graphing a linear equation is the mathematical equivalent to connecting the dots. The x and y values for a linear equation are simply points along a straight line that you can plot on a graph. Chapter 4 shows you how to glean helpful information out of an equation and reveals a number of useful techniques for plotting equations on a graph.

Graphing Linear Equations

In this Chapter...

Introduction to the Coordinate Plane

The coordinate plane is a grid used to illustrate equations. The grid is divided by two lines. One line is horizontal and is known as the x-axis, while the other line is vertical and is known as the y-axis. The two lines split the grid into four quarters, called quadrants, which are labeled using the Roman numerals I, II, III and IV. The location at which the axes meet is called the origin.

A coordinate is a number representing the position of a point with respect to an axis. The combination of an *x* coordinate and a *y* coordinate represents an intersection in the coordinate plane and is called an ordered pair, or coordinate pair. An ordered pair is enclosed in parentheses, with the *x* coordinate listed first, followed by the *y* coordinate.

With an ordered pair, you can easily find the corresponding point in the coordinate plane. For example, the ordered pair (1,–3) represents 1 move right of the origin and 3 moves down from the x-axis.

About the Coordinate Plane

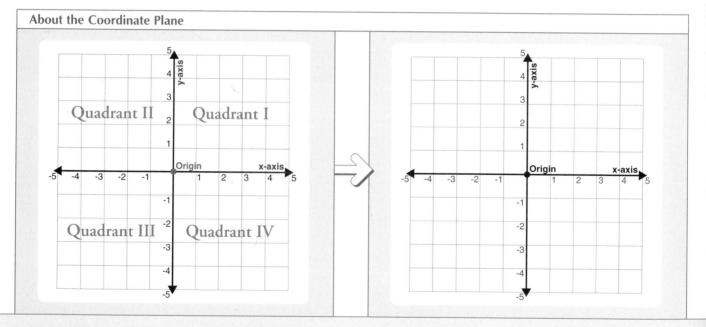

- The coordinate plane contains a horizontal line, called the x-axis, and a vertical line, called the y-axis.

- The x-axis and y-axis intersect at a point called the origin and divide the coordinate plane into four sections, called quadrants.

- The four quadrants are labeled with the Roman numerals I, II, III and IV, starting with the top right quadrant and moving counterclockwise.

- On the x-axis, the numbers to the right of the origin are positive and the numbers to the left of the origin are negative.

- On the y-axis, the numbers above the origin are positive and the numbers below the origin are negative.

Tip

How can I quickly determine which quadrant a point is in?

You can quickly get an idea about where a point lies in the coordinate plane from the point's *x* and *y* values. If the *x* and *y* coordinates are both positive, the point will be in quadrant I. If the *x* coordinate is negative and the *y* coordinate is positive, the point will be in quadrant II. If the *x* and *y* coordinates are both negative, the point will be in quadrant III. If the *x* coordinate is positive and the *y* coordinate is negative, the point will be in quadrant IV.

Tip

Can a coordinate plane accommodate really large or small numbers?

You can change a coordinate plane to accommodate large or small numbers. Each line along the x and y axes can be made to represent 1, 5, 10, 100 or any number of units. Although the lines are generally labeled with integers, they can also be marked with fractions, as long as the fractions increase at a consistent rate. Both axes must use the same units to avoid distorting the lines.

Ordered Pairs

X or Y Coordinate Values of Zero

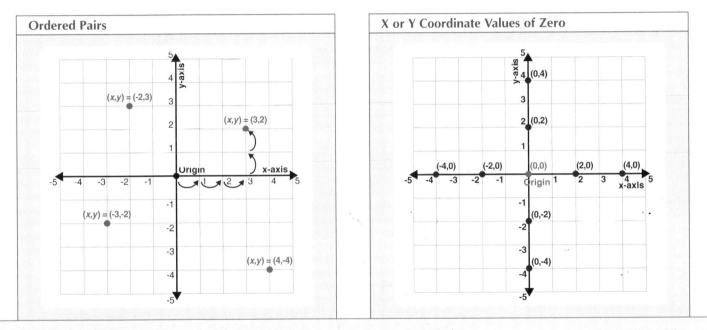

- An ordered pair is two numbers, called coordinates, written as (x,y), that can give the location of any point in the coordinate plane.

- The first number in an ordered pair, known as the *x* coordinate, tells you how far to the left or right of the origin a point is located along the x-axis.

- The second number in an ordered pair, known as the *y* coordinate, tells you how far above or below the origin a point is located along the y-axis.

- The ordered pair for the origin is (0,0).

- When the *x* coordinate in an ordered pair is 0, such as (0,2) or (0,–2), the point is located on the y-axis.

- When the *y* coordinate in an ordered pair is 0, such as (2,0) or (–2,0), the point is located on the x-axis.

CONTINUED

Introduction to the Coordinate Plane *continued*

Creating a coordinate plane is easy. You can draw one by sketching out two perpendicular lines that will represent the x and y axes. Then all you have to do is mark your numbers along each axis. You can sketch gridlines to help find your coordinates more easily or leave the gridlines out, if you prefer. Creating a coordinate plane is a snap if you have a sheet of graph paper handy.

Finding a point in the coordinate plane is called plotting a point. To plot a point, start by moving from the origin, or intersection of the two axes, along the x-axis until you reach the x coordinate specified by the ordered pair. From that point, move along the y-axis until you reach the y coordinate. The points you plot do not have to line up with the lines on your grid. If the numbers are between your gridlines, just estimate their location.

Plot Points

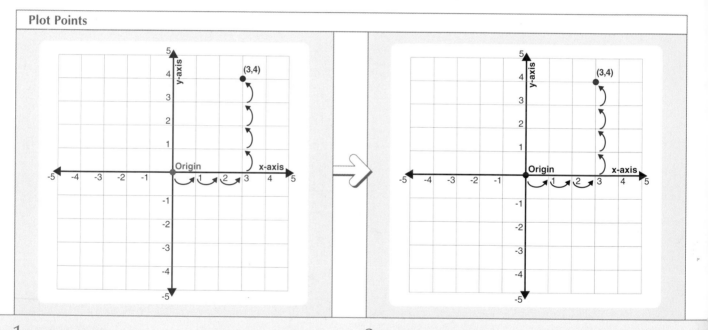

1 To plot, or mark, a point in the coordinate plane, start at the origin and move along the x-axis until you reach the x coordinate. Move to the right for positive numbers and to the left for negative numbers.

- For example, the point (3,4) is three moves to the right of the origin on the x-axis.

  Note: If the x coordinate is zero, do not move from the origin.

2 From the location of the x coordinate, move up or down until you reach the y coordinate. Move up for positive numbers and down for negative numbers.

- For example, the point (3,4) is four moves up along the y-axis.

3 Plot, or mark, the point.

  Note: If the y coordinate is zero, plot the point on the x-axis.

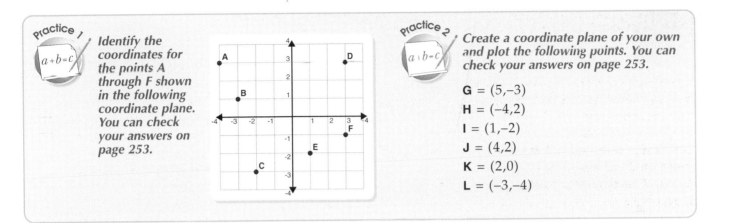

Practice 1

$a+b=c$

Identify the coordinates for the points A through F shown in the following coordinate plane. You can check your answers on page 253.

Practice 2

$a \cdot b=c$

Create a coordinate plane of your own and plot the following points. You can check your answers on page 253.

G = (5,–3)

H = (–4,2)

I = (1,–2)

J = (4,2)

K = (2,0)

L = (–3,–4)

Plotting Points Whose Coordinates are Fractions

The Coordinate Plane Without Gridlines

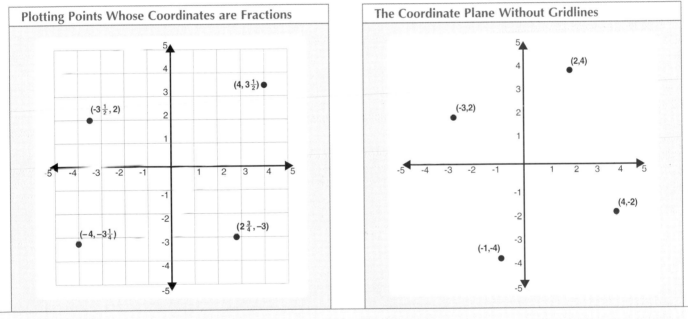

- When plotting points in the coordinate plane, you can also plot points whose coordinates are fractions. For example, you can plot $(4,3\frac{1}{2})$.

- If the lines along the axes are labeled with integers, when you plot coordinates which are fractions, you plot the points in the spaces between the lines in the coordinate plane.

- For example, the point $(4,3\frac{1}{2})$ is four moves to the right on the x-axis and then three and a half moves up along the y-axis.

- A coordinate plane contains a horizontal line for the x-axis and a vertical line for the y-axis. A coordinate plane, however, can be shown without gridlines.

- You can draw a coordinate plane with or without gridlines, depending on your personal preference. In some cases, you may find the coordinate plane easier to work with when you leave out the gridlines.

Graph a Line

All of those equations with x's and y's are more than just numbers and letters—they represent lines. Graphing a line provides an at-a-glance representation of all the solutions, including fractions, to an equation. The values for x and y that solve an equation provide the ordered pairs that you plot in a coordinate plane.

Equations that result in straight lines are called linear equations. You can easily spot a linear

equation because the terms in a linear equation will only contain variables with exponents of 1, such as $2x + 4y = 10$. When graphing a linear equation, you plot points and connect the points with a straight line. You only need to plot two points to draw a line, but plotting a third point will help you make sure that you haven't made a mistake in your calculations.

Step 1: Solve the Equation for One Variable

$$2x + y = 7$$
$$2x + y - 2x = 7 - 2x$$
$$y = 7 - 2x$$

Step 2: Create Ordered Pairs

Let $x = 1$
$$y = 7 - 2x$$
$$y = 7 - 2(1)$$
$$y = 7 - 2$$
$$y = 5$$

Ordered pair $(x,y) = (1,5)$

- You will first need to solve the equation for one of the variables, such as y.

1 Determine which numbers and variables you need to add, subtract, multiply or divide by to place the variable you are solving for by itself on one side of the equation, usually the left side.

2 Add, subtract, multiply or divide both sides of the equation by the numbers and/or variables you determined in step 1.

- In this example, subtract $2x$ from both sides of the equation.

Note: For information on solving equations with more than one variable, see page 70.

3 Choose a random number for the other variable, such as x. For example, let x equal 1.

4 Place the random number you selected into the equation to determine the value of the other variable. Then solve the equation.

5 Write the x and y value together as an ordered pair in the form (x,y).

Note: An ordered pair is two numbers written as (x,y), which gives the location of a point in the coordinate plane.

Tip

How do I graph a linear equation that has only one variable?

When you have an equation with x as the only variable, simply draw a vertical line. For example, for the equation $x = 4$, the x value will always be the same (4) and you can use any number for the y value in the ordered pairs, such as (4,1), (4,2), (4,3), (4,4). Similarly, an equation that only has y as a variable is represented by a horizontal line. For example, for the equation $y = 2$, the y value will always be the same (2) and you can use any number for the x value in the ordered pairs, such as (1,2), (2,2), (3,2), (4,2).

Practice

$a + b = c$

Graph the following lines in their own coordinate planes. You can check your answers on page 253.

1) $y = x$

2) $y = -x$

3) $x = y - 5$

4) $y = 2x + 3$

5) $2x = 3y - 4$

6) $2x - y = 3$

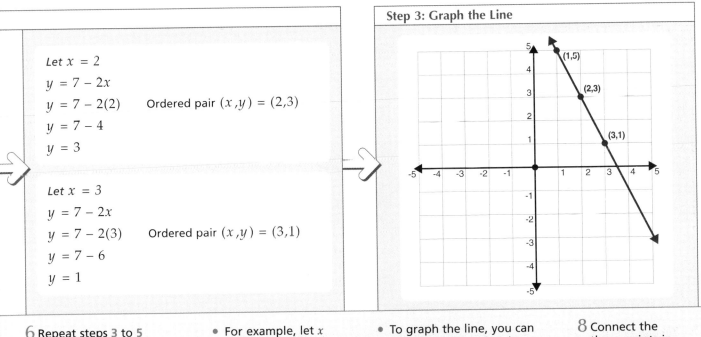

Let $x = 2$

$y = 7 - 2x$

$y = 7 - 2(2)$ Ordered pair $(x,y) = (2,3)$

$y = 7 - 4$

$y = 3$

Let $x = 3$

$y = 7 - 2x$

$y = 7 - 2(3)$ Ordered pair $(x,y) = (3,1)$

$y = 7 - 6$

$y = 1$

Step 3: Graph the Line

6 Repeat steps 3 to 5 twice to determine two more ordered pairs.

● For example, let x equal 2 and let x equal 3.

● To graph the line, you can plot the three points in a coordinate plane to show all the possible solutions to the equation.

7 Plot the three points in a coordinate plane.

Note: For information on plotting points in a coordinate plane, see page 82.

8 Connect the three points in a straight line.

9 Draw an arrow at each end of the line to show that the line extends forever in each direction.

Find the Intercepts of a Line

Despite what Mom or Dad told you, taking the easy way is not always wrong—sometimes it just makes sense. Finding the intercepts of a line is an easy and sensible way to graph a linear equation.

Almost every line which represents a linear equation crosses both the x and y axes. Finding the points, called intercepts, where a line crosses each axis will give you two points you can use to draw a straight line and graph the equation.

Intercepts are easier to plot than other points along the line because finding the intercepts

involves using that magical multiplier of 0. The x-intercept always takes the form (x,0), so you simply have to plug y = 0 into the equation to find x. Similarly, because the y-intercept always takes the form (0,y), you just plug x = 0 into the equation to find y.

Once you have those two ordered pairs, plot them in the coordinate plane and connect the dots with a ruler. Then you have a line representing the equation and are done in no time flat.

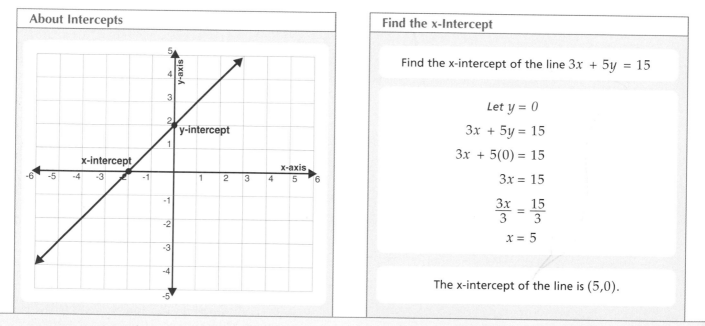

About Intercepts

Find the x-Intercept

Find the x-intercept of the line $3x + 5y = 15$

$$Let\ y = 0$$

$$3x + 5y = 15$$

$$3x + 5(0) = 15$$

$$3x = 15$$

$$\frac{3x}{3} = \frac{15}{3}$$

$$x = 5$$

The x-intercept of the line is (5,0).

- An intercept is a point where a line crosses either the x-axis or the y-axis in the coordinate plane.

  Note: For information on the coordinate plane, see page 80.

- The x-intercept is the point where a line crosses the x-axis.

- The y-intercept is the point where a line crosses the y-axis.

- To find the x-intercept of a line, you need to let y equal zero and solve for x in the equation.

1 In the equation, replace y with 0 and solve the equation to find the value of x.

  Note: For information on solving equations, see pages 64 to 67.

2 Write the x and y values together as an ordered pair in the form (x,y). Every x-intercept has a y value of zero, such as (5,0).

  Note: An ordered pair is two numbers written as (x,y), which gives the location of a point in the coordinate plane.

Tip

Do some lines have only one intercept?

Yes. A vertical line only crosses the x-axis, so the line has only an x-intercept. Likewise, a horizontal line only crosses the y-axis, so the line has only a y-intercept.

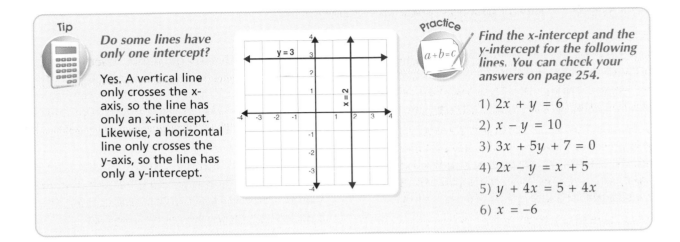

Practice

$a+b=c$

Find the x-intercept and the y-intercept for the following lines. You can check your answers on page 254.

1) $2x + y = 6$

2) $x - y = 10$

3) $3x + 5y + 7 = 0$

4) $2x - y = x + 5$

5) $y + 4x = 5 + 4x$

6) $x = -6$

Find the y-Intercept

Find the y-intercept of the line $3x + 5y = 15$

Let $x = 0$

$3x + 5y = 15$

$3(0) + 5y = 15$

$5y = 15$

$\dfrac{5y}{5} = \dfrac{15}{5}$

$y = 3$

The y-intercept of the line is $(0,3)$.

Graph a Line

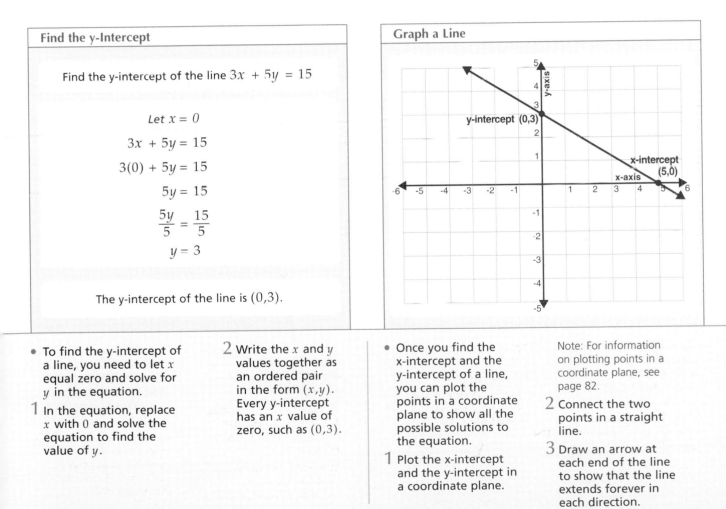

- To find the y-intercept of a line, you need to let x equal zero and solve for y in the equation.

1 In the equation, replace x with 0 and solve the equation to find the value of y.

2 Write the x and y values together as an ordered pair in the form (x,y). Every y-intercept has an x value of zero, such as $(0,3)$.

- Once you find the x-intercept and the y-intercept of a line, you can plot the points in a coordinate plane to show all the possible solutions to the equation.

1 Plot the x-intercept and the y-intercept in a coordinate plane.

Note: For information on plotting points in a coordinate plane, see page 82.

2 Connect the two points in a straight line.

3 Draw an arrow at each end of the line to show that the line extends forever in each direction.

Find the Slope of a Line

If you have ever been downhill skiing, you probably understand the significance of slope—the steeper the hill, the faster you get to the bottom. Slope is actually a mathematical expression that tells you how much a line slants and in which direction.

Calculating the slope of a line is a fairly easy process. Taking two points along the line, you simply divide how high the line rises on the y-axis by how far the line travels along the x-axis. The resulting value, usually represented by the variable m, will tell you a lot about the line without you having to graph the line. If the slope is positive, the line rises from left to right. If the slope is negative, the line falls from left to right. The farther the value of the slope is from 0, the steeper the line is.

Finding the slope of a line is helpful in tracking statistical trends. For example, you may want to create a line that tracks gas prices over several months. The steeper the slope of the line, the faster gas prices are rising and the faster you will decide to buy a fuel-efficient car.

About the Slope of a Line

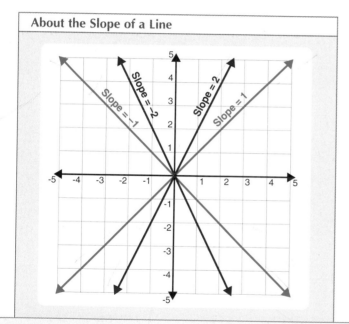

- The slope of a line indicates the steepness and direction of a line.
- If a line's slope is a positive number, the line slants to the right.
- If a line's slope is a negative number, the line slants to the left.

- The further a slope's value is from 0, whether positive or negative, the steeper the line.
- The closer a slope's value is to 0, whether positive or negative, the more flat the line.

Finding the Slope of a Line

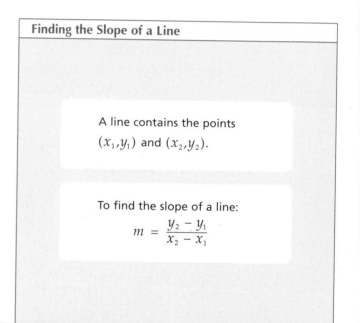

A line contains the points (x_1, y_1) and (x_2, y_2).

To find the slope of a line:

$$m = \frac{y_2 - y_1}{x_2 - x_1}$$

- When you know the x and y coordinates of any two points on a line, you can find the slope of the line, which is represented by the letter m.

Note: For information on x and y coordinates, see page 81.

1 To find the slope of a line, subtract the y values and divide the result by the difference between the x values.

Note: Subscripts, which are the small numbers beside the letters x and y, are used to identify the first (x_1, y_1) and second (x_2, y_2) points in a line.

Tip

What are some of the most common mistakes that people make when they try to determine the slope of a line?

A common mistake is to subtract the *x* values, instead of the *y* values, on the top of the slope formula. The slope always equals $y_2 - y_1$ divided by $x_2 - x_1$, not the other way around. Another common mistake occurs when people mix up the order of the points in the slope formula. You should decide which point is first and which point is second and then be sure to subtract the first *y* value from the second, and the first *x* value from the second.

Practice

$a + b = c$

Find the slope of the line that passes through the following points. You can check your answers on page 254.

1) (0,0) and (3,7)

2) (1,2) and (−1,3)

3) (−1,0) and (0,2)

4) (3,−6) and (5,−10)

5) (2,2) and (3,3)

6) (2,4) and (9,4)

Finding the Slope of a Line Example

$$m = \frac{y_2 - y_1}{x_2 - x_1}$$

Find the slope of the line that passes through (1,2) and (3,6).

Let (1,2) be (x_1, y_1) and (3,6) be (x_2, y_2).

$$m = \frac{6 - 2}{3 - 1} = \frac{4}{2} = 2$$

- In this example, the slope of a line is determined by subtracting the *y* values (6 − 2) and then dividing the result by the difference between the *x* values (3 − 1). The slope of the line is 2.

Note: You will end up with the same slope if you subtract the *x* and *y* values in the opposite order. In this example, you could subtract the *y* values (2 − 6) and then divide the result by the difference between the *x* values (1 − 3). This will also give you a slope of 2.

Finding the Slope of a Horizontal or Vertical Line

$$m = \frac{y_2 - y_1}{x_2 - x_1}$$

Find the slope of the horizontal line that passes through (1,4) and (5,4).

Let (1,4) be (x_1, y_1) and (5,4) be (x_2, y_2).

$$m = \frac{4 - 4}{5 - 1} = \frac{0}{4} = 0$$

Find the slope of the vertical line that passes through (3,1) and (3,2).

Let (3,1) be (x_1, y_1) and (3,2) be (x_2, y_2).

$$m = \frac{2 - 1}{3 - 3} = \frac{1}{0} = \text{undefined}$$

- A horizontal line will always have a slope of 0. Two points on a horizontal line will always have the same *y* value, so the top number in the slope formula will always be 0 when you find the slope of the line.

- A vertical line does not have a slope and its slope is considered undefined. Two points on a vertical line will always have the same *x* value, so the bottom number in the slope formula will always be 0 when you find the slope of the line. You cannot divide a number by 0.

Write an Equation in Slope-Intercept Form

Don't you love it when something familiar gets a fancy new name? A classic example has garbage collectors renamed "sanitary engineers." Well, get ready for another term like that. When you have solved for y in an equation, you can say that the equation is expressed in "slope-intercept form."

As you have learned, there are many ways to write an equation. When you solve any equation for y, however, you automatically receive a bunch of useful information about a line—namely the slope of the line and the y-intercept. In mathspeak,

slope-intercept form is written as $y = mx + b$, where m is the slope of the line and b is the y-intercept. Therefore, in the example $y = 3x - 7$, you can see that the slope of the line is 3. Since you know that the x coordinate of a y-intercept is 0, you can also see that the y-intercept is $(0,-7)$.

Writing an equation in slope-intercept form comes in handy because it allows you to draw a line in the coordinate plane with relative ease. For information on graphing a line using slope-intercept form, see page 92.

About Slope-Intercept Form

$y = 3x + 5$ The slope is 3.
The y-intercept is $(0,5)$.

$y = 2x - 4$ The slope is 2.
The y-intercept is $(0,-4)$.

$y = -6x + 2$ The slope is −6.
The y-intercept is $(0,2)$.

$y = \frac{1}{4}x - 7$ The slope is $\frac{1}{4}$.
The y-intercept is $(0,-7)$.

Write an Equation in Slope-Intercept Form

$$4x + 2y = 6$$
$$4x + 2y - 4x = 6 - 4x$$
$$2y = 6 - 4x$$

- When the equation of a line is written in slope-intercept form, you can immediately identify the slope of the line and the y-intercept of the line.

- The number in front of x is the slope of the line, which indicates the steepness and direction of the line.

- The number on its own is the y-intercept, which is the point where a line crosses the y-axis. Every y-intercept has an x value of zero, such as $(0,5)$.

Note: For more information on the slopes of lines, see page 88. For more information on intercepts, see page 86.

- You can write an equation in slope-intercept form by placing the y variable by itself on the left side of the equation.

1 Determine which numbers and variables you need to add and subtract to place the y variable by itself on the left side of the equation.

2 Add and/or subtract the numbers and/or variables you determined in step 1 on both sides of the equation.

- In this example, subtract $4x$ from both sides of the equation.

Note: For information on adding and subtracting numbers and variables in equations, see page 64.

Tip

An equation does not have an x variable. How do I determine the slope of the line?

The equation of a horizontal line, such as $y = 8$, does not include an x variable. To find the slope of the horizontal line, you may want to write the equation in slope-intercept form, such as $y = 0x + 8$. Writing the equation in this format clearly shows that the slope of the line is 0.

Practice

$a + b = c$

Write the following equations in slope-intercept form and identify the slope and y-intercept of each line. You can check your answers on page 254.

1) $y - x = 8$

2) $2y + 6x + 10 = 0$

3) $4 = x - 3y$

4) $2x + y = x + 4$

5) $x + y = x - y$

6) $3x - 2y + 3 = 2x + 3y + 2$

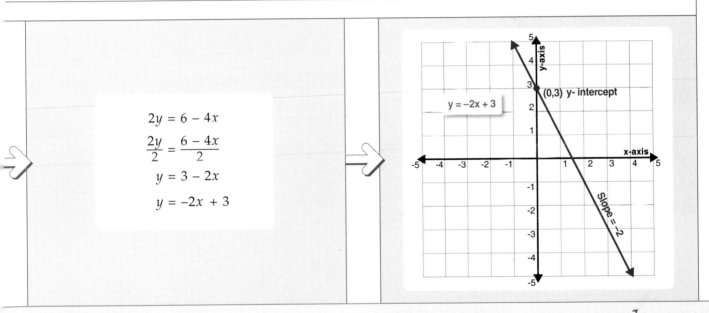

$$2y = 6 - 4x$$
$$\frac{2y}{2} = \frac{6 - 4x}{2}$$
$$y = 3 - 2x$$
$$y = -2x + 3$$

$y = -2x + 3$

(0,3) y- intercept

Slope = -2

3 Determine which numbers you need to multiply or divide to place the y variable by itself on the left side of the equation.

4 Multiply or divide both sides of the equation by the numbers you determined in step **3**.

Note: For information on multiplying and dividing numbers in equations, see page 66.

- In this example, divide by 2 on both sides of the equation.

Note: You can rearrange the terms in an equation so the x variable appears before the number on the right side of the equation. For example, you can write $y = 3 - 2x$ as $y = -2x + 3$.

- Once you write an equation in slope-intercept form, you can easily determine the slope and the y-intercept of the line.

- In this example, the slope of the line is -2 and the y-intercept is (0,3).

Note: For information on graphing a line, see page 84.

Graph a Line in Slope-Intercept Form

A linear equation presented in slope-intercept form is a fancy way of saying that the equation has been solved for *y*. No matter what you call it, though, you're about to find out that a linear equation in slope-intercept form offers one of the easiest ways to graph a line.

Once you have a linear equation expressed in slope-intercept form, determine the slope and the y-intercept of the line. This is a snap because they can be found on the right side of the equation. For example, if you have an equation of $y = 4x - 7$, the slope is 4 and the y-intercept is $(0,-7)$.

Next, plot the y-intercept in the coordinate plane. From the location of the y-intercept, you can use the slope to count your way to a second point on the line. Remember that the slope tells you how high the line rises on the y-axis and how far the line travels along the x-axis. Using a slope of 4, which can also be expressed as $\frac{4}{1}$, you would count up 4 units and right 1 unit, landing on the location $(1,-3)$ in the coordinate plane.

By simply connecting the two points with a ruler, you have a line which represents the equation.

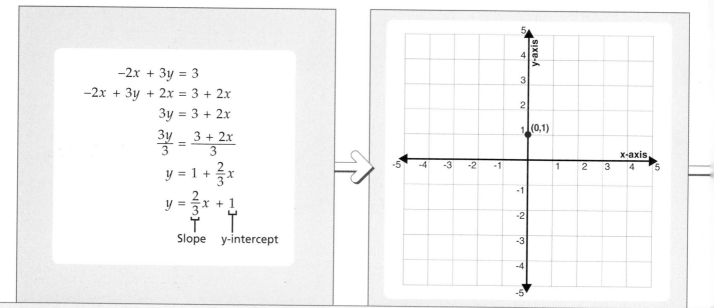

$$-2x + 3y = 3$$
$$-2x + 3y + 2x = 3 + 2x$$
$$3y = 3 + 2x$$
$$\frac{3y}{3} = \frac{3 + 2x}{3}$$
$$y = 1 + \frac{2}{3}x$$
$$y = \underset{\text{Slope}}{\frac{2}{3}}x + \underset{\text{y-intercept}}{1}$$

1 Write the equation you want to graph in slope-intercept form. Once you write an equation in slope-intercept form, you can easily determine the slope of the line and the y-intercept of the line.

2 Determine the slope and y-intercept of the line. A y-intercept always has an *x* value of zero, such as $(0,1)$.

• In this example, the slope of the line is $\frac{2}{3}$ and the y-intercept is $(0,1)$.

Note: To write an equation in slope-intercept form and determine the slope and y-intercept of a line, see page 90.

3 Plot the y-intercept in a coordinate plane. The y-intercept is the point where a line crosses the y-axis.

Note: For information on plotting points in a coordinate plane, see page 82.

Tip

What should I consider when using the slope of a line to graph the line?

If the slope of a line is a whole number, you will need to write the number as a fraction to graph the line. Remember that all whole numbers have an invisible denominator of 1. If the slope of a line is a negative number, place a negative sign (–) in the top or bottom part of the fraction—it does not matter which part since the resulting line will be the same. For example, if the slope of the line is $-\frac{2}{3}$, write the fraction as $\frac{-2}{3}$ or as $\frac{2}{-3}$. Just be sure not to add the negative sign to both numbers, such as $\frac{-2}{-3}$, as this will change the line you draw.

Practice

Graph the following equations using the slope-intercept form of a line. You can check your graphs on page 255.

1) $y = 3x + 1$

2) $y = -x + 5$

3) $y = x - 2$

4) $y = -2x + 4$

5) $y = 3$

6) $y = -x - 3$

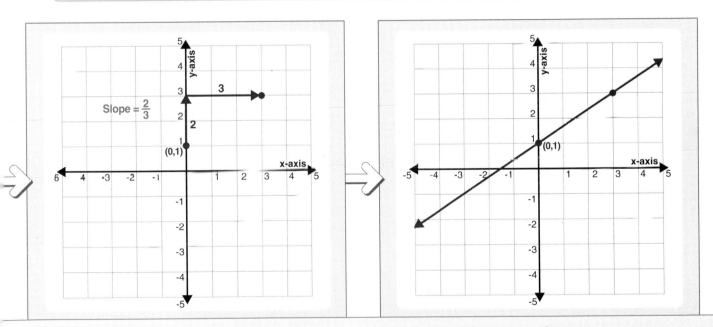

4 To plot another point using the slope of the line, start at the y-intercept and move along the y-axis the same number of units as the numerator, or top number, of the slope. Move up for a positive number and down for a negative number.

5 From your final location in step **4**, move along the x-axis the same number of units as the denominator, or bottom number, of the slope. Move to the right for a positive number and to the left for a negative number.

6 Plot the point.

7 Connect the two points in a straight line.

8 Draw an arrow at each end of the line to show that the line extends forever in each direction. The line shows all the possible solutions to the equation.

Write an Equation in Point-Slope Form

The point-slope formula allows you to figure out a linear equation given nothing more than one point along a line and the slope of the line.

The point-slope formula is expressed as $y - y_1 = m(x - x_1)$. While this formula may look complicated, it is actually pretty simple—x_1 and y_1 represent the coordinates of a point on the line and m represents the slope of the line. For example, given the point (4,5) and a slope of 2, you would create the equation $y - 5 = 2(x - 4)$.

In addition, when you encounter linear equations expressed in the point-slope form, you will be able to immediately determine a point on a line and the slope of the line.

After creating a linear equation in point-slope form, you may sometimes be asked to write the equation in slope-intercept form, which is really nothing more than simplifying and reorganizing the equation. For example, the equation $y - 5 = 2(x - 4)$ is the same as the equation $y = 2x - 3$. For more information on slope-intercept form, see page 90.

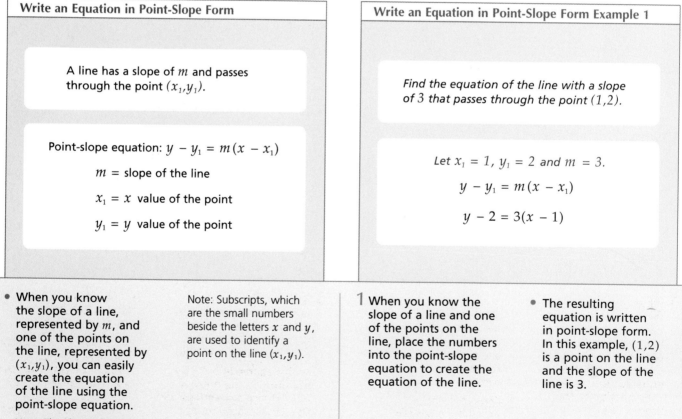

Write an Equation in Point-Slope Form

A line has a slope of m and passes through the point (x_1, y_1).

Point-slope equation: $y - y_1 = m(x - x_1)$

m = slope of the line

$x_1 = x$ value of the point

$y_1 = y$ value of the point

Write an Equation in Point-Slope Form Example 1

Find the equation of the line with a slope of 3 that passes through the point (1,2).

Let $x_1 = 1$, $y_1 = 2$ and $m = 3$.

$y - y_1 = m(x - x_1)$

$y - 2 = 3(x - 1)$

- **When you know the slope of a line, represented by m, and one of the points on the line, represented by (x_1, y_1), you can easily create the equation of the line using the point-slope equation.**

  Note: The slope of a line indicates the steepness and direction of a line.

Note: Subscripts, which are the small numbers beside the letters x and y, are used to identify a point on the line (x_1, y_1).

1 When you know the slope of a line and one of the points on the line, place the numbers into the point-slope equation to create the equation of the line.

- The resulting equation is written in point-slope form. In this example, (1,2) is a point on the line and the slope of the line is 3.

Tip

What can I do if I am only given two points of a line and asked to express the line as an equation?

To write an equation for the line, you must first use the two given points to determine the slope of the line (see page 88). Once you have determined the slope of the line, you can then place the slope and one of the given points into the point-slope equation to write the equation of the line.

Practice

Find the equation of the following lines using the slope and point given for each line. Write the equations in point-slope form. You can check your answers on page 255.

1) slope = 2, point = (1,3)

2) slope = −3, point = (0,4)

3) slope = −1, point = (1,−2)

4) slope = 5, point = (−3,7)

5) slope = 0, point = (1,1)

6) slope = 12, point = (−3,−6)

Write an Equation in Point-Slope Form Example 2

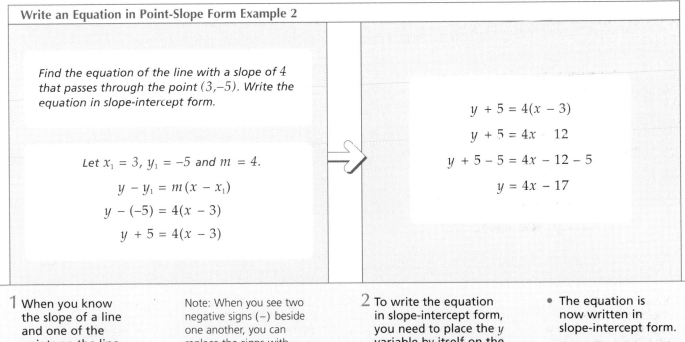

Find the equation of the line with a slope of 4 that passes through the point $(3,-5)$. Write the equation in slope-intercept form.

Let $x_1 = 3$, $y_1 = -5$ and $m = 4$.

$$y - y_1 = m(x - x_1)$$

$$y - (-5) = 4(x - 3)$$

$$y + 5 = 4(x - 3)$$

$$y + 5 = 4(x - 3)$$

$$y + 5 = 4x - 12$$

$$y + 5 - 5 = 4x - 12 - 5$$

$$y = 4x - 17$$

1 When you know the slope of a line and one of the points on the line, place the numbers into the point-slope equation to create the equation of the line.

Note: When you see two negative signs (−) beside one another, you can replace the signs with one positive sign (+). For example, $y - (-5)$ becomes $y + 5$.

- The resulting equation is written in point-slope form.

2 To write the equation in slope-intercept form, you need to place the y variable by itself on the left side of the equation.

- In this example, start by multiplying the number outside the parentheses by the variable and number inside the parentheses to remove the parentheses ( ). Then subtract 5 from both sides of the equation.

- The equation is now written in slope-intercept form.

Note: When the equation of a line is written in slope-intercept form, you can immediately identify the slope of the line and the y-intercept of the line. For more information on the slope-intercept form, see page 90.

Parallel and Perpendicular Lines

Parallel lines are like railway tracks. The tracks travel beside one another, remaining exactly the same distance apart and never touching each other. Due to the fact that parallel lines never cross, the lines do not share a common point in the coordinate plane. Parallel lines also have exactly the same slope, which indicates the steepness and direction of a line.

Unlike parallel lines, perpendicular lines do cross one another, at perfect 90-degree angles. The slopes of two perpendicular lines are negative reciprocals of one another. For example, if the slope of a line is $\frac{2}{3}$, the slope of a perpendicular line will be $-\frac{3}{2}$.

Just by looking at the slopes of two lines, you can easily tell if the lines are parallel or perpendicular.

Keep in mind, however, that some lines may throw you a curve ball. A horizontal line will always have a slope of 0 and the negative reciprocal of 0 is undefined. The slope of a vertical line is considered undefined, so you cannot find the negative reciprocal of the slope of a vertical line either.

About Parallel Lines

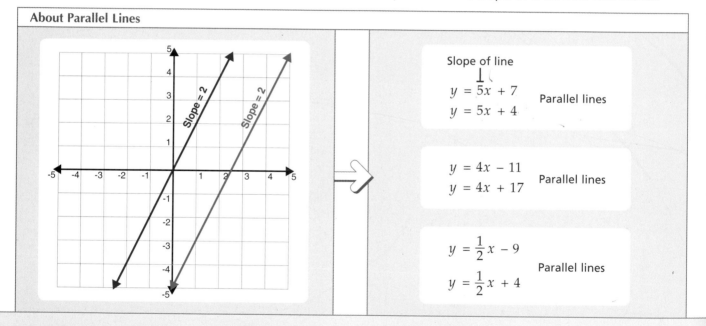

Slope of line
$$y = 5x + 7$$
$$y = 5x + 4$$
Parallel lines

$$y = 4x - 11$$
$$y = 4x + 17$$
Parallel lines

$$y = \frac{1}{2}x - 9$$
$$y = \frac{1}{2}x + 4$$
Parallel lines

- Parallel lines never intersect.
- Parallel lines have the same slope, which always keeps the lines the same distance apart from one another.

Note: The slope of a line indicates the steepness and direction of a line.

- When the equation of a line is written in slope-intercept form, you can quickly identify the slope of the line. The slope of the line appears in front of the x variable.

Note: For information on writing an equation in slope-intercept form, see page 90.

- When two lines have the same slope, the lines are parallel.

Tip

How do I find the negative reciprocal of a number?

Finding the negative reciprocal of a fraction is easy—just flip it over, switching the top and bottom numbers, and then change the sign (+ or –) in front of the fraction. For instance, the negative reciprocal of $\frac{3}{4}$ is $-\frac{4}{3}$. To find the negative reciprocal of a whole number, write the number as a fraction, keeping in mind that all whole numbers have an invisible denominator of 1. Then flip the fraction and change the sign (+ or –) in front of the fraction. For example, the number 3 can be written as $\frac{3}{1}$, so its negative reciprocal is $-\frac{1}{3}$.

Practice

$a + b = c$

State whether the following lines are parallel or perpendicular. For some equations, you may first need to write the equations in slope-intercept form (see page 90). You can check your answers on page 255.

1) $y = 3x - 4$, $y = 3x + 10$

2) $y = x + 2$, $y = -x - 4$

3) $y = \frac{2x}{5} - 1$, $y = -\frac{5x}{2}$

4) $y - 1 = 4(x - 2)$, $y = 4x - 3$

5) $2x - y = 6 - x$, $y + x - 1 = -y + 7x$

6) $y = 3x - 1$, $y = -4x + 5$

About Perpendicular Lines

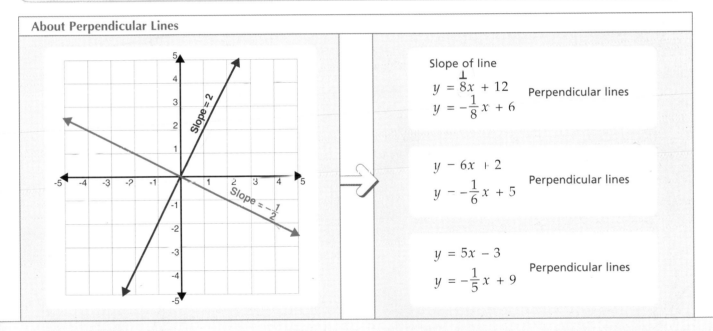

Slope of line
⊥
$y = 8x + 12$
$y = -\frac{1}{8}x + 6$ Perpendicular lines

$y = 6x + 2$
$y = -\frac{1}{6}x + 5$ Perpendicular lines

$y = 5x - 3$
$y = -\frac{1}{5}x + 9$ Perpendicular lines

- Perpendicular lines intersect one another at right angles, or 90-degree angles.

- Perpendicular lines have slopes that are negative reciprocals of one another. For example, 2 is the negative reciprocal of $-\frac{1}{2}$.

Note: For more information on negative reciprocals, see the top of this page.

- When the equation of a line is written in slope-intercept form, you can quickly identify the slope of the line. The slope of the line appears in front of the x variable.

- When two lines have slopes which are negative reciprocals of one another, the lines are perpendicular.

CONTINUED

Parallel and Perpendicular Lines *continued*

As you have seen, parallel lines are lines that have the same slope and travel beside each other without touching. Remember also that perpendicular lines are two lines that intersect at 90-degree angles and have slopes that are the negative reciprocal of one another.

In algebra, you will sometimes be given a line and asked to create an equation for a line that is either parallel or perpendicular to the original line. Understanding the unique traits of both parallel and perpendicular lines is crucial if you want to solve these problems.

You will also need to brush up on your knowledge of the slope-intercept and point-slope forms of linear equations. An equation in slope-intercept form gives you the slope of a line and its y-intercept. An equation in point-slope form gives you the slope of a line along with one point on the line.

Find the Equation of a Line

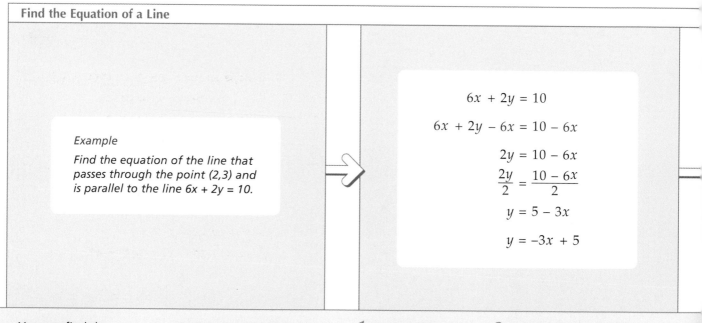

Example

Find the equation of the line that passes through the point (2,3) and is parallel to the line 6x + 2y = 10.

$$6x + 2y = 10$$
$$6x + 2y - 6x = 10 - 6x$$
$$2y = 10 - 6x$$
$$\frac{2y}{2} = \frac{10 - 6x}{2}$$
$$y = 5 - 3x$$
$$y = -3x + 5$$

- You can find the equation of a line that passes through a specific point and is parallel or perpendicular to another line.

- To find the equation of a line, you may first need to write the equation of the line you are given in slope-intercept form.

  Note: For more information on the slope-intercept form of a line, see page 90.

1 To write the equation of the line you are given in slope-intercept form, determine which numbers and/or variables you need to add, subtract, multiply or divide to place the y variable by itself on the left side of the equation.

2 Add, subtract, multiply and/or divide both sides of the equation by the numbers and variables you determined in step 1.

- In this example, subtract $6x$ from both sides of the equation and then divide both sides of the equation by 2.

- The equation is now written in slope-intercept form.

Practice $a + b = c$

For each question, find the equation of the line that passes through the given point and is parallel to the given line. You can check your answers on page 256.

1) Point: (0,0)

Equation: $y = 2x - 1$

2) Point: (−1,2)

Equation: $y = -5x + 7$

3) Point: (2,3)

Equation: $y - x + 1 = 2x + 4$

Practice $a + b = c$

For each question, find the equation of the line that passes through the given point and is perpendicular to the given line. You can check your answers on page 256.

a) Point: (0,1)

Equation: $y = -2x + 5$

b) Point: (−2,−5)

Equation: $y = \dfrac{3x}{5} - 2$

c) Point: (1,3)

Equation: $-2x - 3 - y = y + 4x + 1$

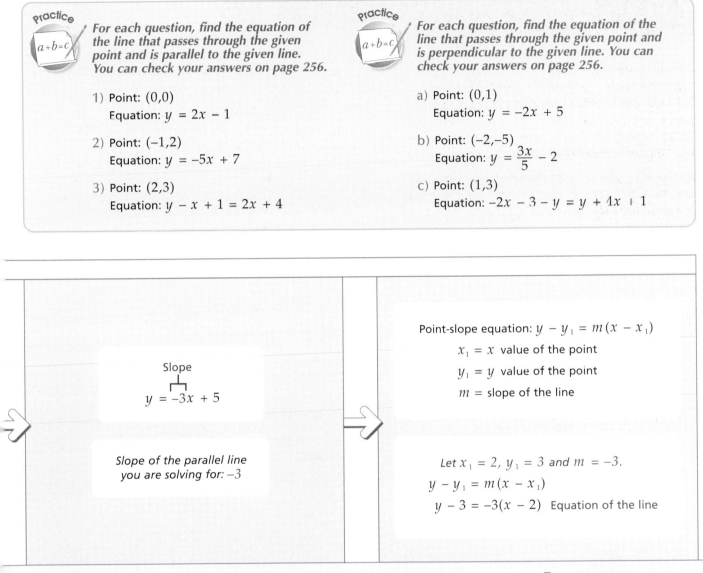

Slope

$y = -3x + 5$

Slope of the parallel line
you are solving for: −3

Point-slope equation: $y - y_1 = m(x - x_1)$

$x_1 = x$ value of the point

$y_1 = y$ value of the point

$m =$ slope of the line

Let $x_1 = 2$, $y_1 = 3$ and $m = -3$.

$y - y_1 = m(x - x_1)$

$y - 3 = -3(x - 2)$ Equation of the line

- When the equation of a line is written in slope-intercept form, you can immediately identify the slope of the line.

3 Determine the slope of the line, which appears in front of the x variable. In this example, the slope of the line is −3.

4 Determine the slope of the line you are solving for. If the lines are parallel, the slopes will be the same. If the lines are perpendicular, the slopes will be negative reciprocals of one another.

- In this example, the lines are parallel, so the slope of the line you are solving for is −3.

- When you know one of the points on a line and the slope of a line, you can use the point-slope equation to create the equation of the line.

Note: For more information on the point-slope equation, see page 94.

5 Place the point on the line you were given and the slope you determined in step 4 into the point-slope equation to create the equation of the line.

- You have found the equation of the line.

99

Graphing Linear Equations

Question 1. Graph the following lines.

a) $y = 2$

b) $x = -1$

c) $y = x + 2$

d) $y = 3 - x$

e) $2x + 4y = 1$

f) $x - y = 0$

g) $x = 4y + 2$

Question 2. Find the x- and y-intercepts of the following lines.

a) $y = 2x - 5$

b) $y = -3x + 6$

c) $x - y = 1$

d) $5x + 7y = 0$

e) $2x + 7y + 1 = 0$

f) $-3x + 2y = 12$

Question 3. Find the slope of the line that passes through the following points.

a) $(1,-7)$ and $(0,-10)$

b) $(0,2)$ and $(2,-6)$

c) $(3,5)$ and $(-6,5)$

d) $(0,-7)$ and $(0,10)$

e) $(2,-11)$ and $(7,-6)$

f) $(0,2)$ and $(5,0)$

Question 4. Write the following equations in slope-intercept form and identify the slope and y-intercept of each line.

a) $y + x + 2 = 0$

b) $2y - 3x = 5$

c) $3x - 4y = 24$

d) $x - y - 1 = 0$

e) $-y + 5x = 2$

f) $3y + 8 - 2x = 0$

Question 5. Determine the equation of the following lines using the slope and point given for each line. Write the equations in slope-intercept form.

a) Slope = 1, point = (3,5)

b) Slope = −2, point = (7,11)

c) Slope = 4, point = (−3, 0)

d) Slope = 6, point = (0,0)

e) Slope = $-\dfrac{1}{2}$, point = (8,−5)

f) Slope = $\dfrac{5}{4}$, point = (4,0)

Question 6. Determine if the following pairs of lines are parallel, perpendicular or neither.

a) $y = 2x + 3$, $y = 2x + 10$

b) $y = -x + 4$, $y = x + 2$

c) $y = 5x$, $y = -\dfrac{x}{5}$

d) $y = 2x + 10$, $y = 3x + 8$

e) $x - y = 5$, $y = x - 2$

f) $2x + y = 7$, $x = 5 + 2y$

You can check your answers on pages 269-271.

Chapter 5

While an equation merely gives you one or a handful of values as a solution, an inequality generally gives you a solution that covers a whole range of values. Fortunately, solving inequalities is very similar to solving equations, but there are a few quirks. Chapter 5 equips you with the skills to solve inequalities and plot them on a graph.

Solving & Graphing Inequalities

In this Chapter...

Introduction to Inequalities

Inequalities are mathematical statements which compare the values of two expressions. In other words, one side of the statement does not exactly equal the other.

Inequalities are similar to equations, but instead of the left and right sides being separated by an equals sign (=), the sides are separated by one of four inequality symbols—greater than (>), less than (<), greater than or equal to (≥), or less than or equal to (≤). For example, 5 < 10 is an inequality. Additionally, while equations provide only one correct answer, inequalities leave room for a variety of answers. For example, if $x > 5$ then x is 6, or 7, or 8, or any number bigger than 5.

When you are working with inequalities, remember that positive numbers are always greater than negative numbers and 0 is less than a positive number but greater than a negative number.

A compound inequality is a slightly more complex statement that shows two inequalities at once. In a compound inequality, the value of a single variable falls within a specific range of values. For example, $-5 < y < 5$ is a compound inequality that defines the value of y as a range of numbers between −5 and 5.

About Inequalities	
2 < 5	2 is less than 5
6 ≤ 20	6 is less than 20 or equal to 20
10 > 7	10 is greater than 7
8 ≥ 3	8 is greater than 3 or equal to 3
15 ≥ 15	15 is greater than 15 or equal to 15

Inequalities: True or False	
50 < 100	true
−6 ≤ −15	false
−30 ≤ −30	true
−20 ≥ 1	false
−2 > −5	true

- An inequality is a mathematical statement in which one side is less than, greater than or possibly equal to the other side.

- Inequalities use four different symbols.

  < less than

  ≤ less than or equal to

  > greater than

  ≥ greater than or equal to

  Note: The point in an inequality symbol (<) points at the smaller number.

- An inequality can be considered true or false.

- When comparing a positive number to a negative number, the positive number is always greater. For example, $-2 < 1$. When comparing two negative numbers, the number closer to zero is always greater. For example, $-2 > -5$.

Practice

$a + b = c$

Describe each of the following inequalities in words. For example, $1 < 5$ would be "one is less than five." You can check your answers on page 256.

1) $-1 < 0$

2) $3 \geq 2$

3) $5 \leq 5$

4) $-1 < x \leq 3$

5) $4 \geq y \geq -3$

6) $0 < 4 < 10$

Practice

$a + b = c$

Determine whether each of the following inequalities is true or false. You can check your answers on page 256.

a) $0 < 4$

b) $3 > 3$

c) $-4 \leq 2$

d) $-1 \geq -1$

e) $-1 < -2$

f) $10 > 5$

Inequalities with Variables

$x < 25$	x is less than 25
$y \leq 30$	y is less than 30 or equal to 30
$a > 7$	a is greater than 7
$b \geq 14$	b is greater than 14 or equal to 14

Compound Inequalities

$8 < x < 14$	x is greater than 8 and less than 14
$40 \leq y < 50$	y is greater than or equal to 40 and less than 50
$15 < a \leq 25$	a is greater than 15 and less than or equal to 25
$10 \leq b \leq 30$	b is greater than or equal to 10 and less than or equal to 30

- An inequality can include a variable. The inequality indicates if a variable is larger, smaller or possibly equal to a specific number.

- For example, $x < 25$ indicates that the value of x can be any value less than 25.

- A compound inequality, such as $8 < x < 14$, defines a range of possible values for a variable. A compound inequality has a lower and upper boundary for the range of possible values.

- When writing a compound inequality, you will usually write the lower boundary on the left and the upper boundary on the right.

Note: If an inequality symbol is <, the lower or upper boundary is not included in the range of possible values. If the inequality symbol is ≤, the lower or upper boundary is included in the range of possible values.

Solve an Inequality with Addition and Subtraction

Good news! If you have done your homework and learned to solve equations with addition and subtraction, you virtually know how to solve inequalities with addition and subtraction already. The only difference between an equation and an inequality is the answer, or more accurately, the answers. Inequalities produce a whole range of answers. For example, if $b > 5$ then $b = 6, 7, 8, 9$, and so on.

When working with inequalities, you must remember to keep the problem balanced.

Whenever you perform an operation on one side of the inequality symbol, you must perform the same operation on the other side so the inequality will remain true.

The object of your adding and subtracting is to "cancel out" numbers and isolate the variable to one side of the inequality—called solving for the variable. For example, in the inequality $y + 5 > 8$ you would subtract 5 from each side of the greater than symbol to cancel out the 5 originally found on the left side. In this example, we would have $y + 5 - 5 > 8 - 5$, which leaves an answer of $y > 3$.

How to Solve an Inequality

$$x - 3 > 8$$
$$x - 3 + 3 > 8 + 3$$
$$x > 11$$

$$12 + x \leq 15$$
$$12 + x - 12 \leq 15 - 12$$
$$x \leq 3$$

Solving Inequalities Example 1

$$x + 4 < 35$$
$$x + 4 - 4 < 35 - 4$$
$$x < 31$$

Check your answer
Let $x = 30$

$x + 4 < 35$
$30 + 4 < 35$
$34 < 35$ *True!*

Check your answer
Let $x = 31$

$x + 4 < 35$
$31 + 4 < 35$
$35 < 35$ *Not true!*

- When solving an inequality, your goal is to place the variable on one side of the inequality, usually the left side, and place all the numbers on the other side of the inequality, usually the right side. This allows you to determine the value of the variable.

- To place a variable by itself on one side of an inequality, you can add or subtract the same number or variable on both sides of the inequality and the inequality will remain true.

1 In this example, to place the variable by itself on the left side of the inequality, subtract 4 from both sides of the inequality.

Note: By subtracting 4 from the left side of the inequality, you effectively cancel out the number 4.

2 To check your answer, choose two numbers that are closest to your answer—one number that should make the inequality true and another number that should make the inequality not true. Place the numbers into the original inequality and solve the problem.

- If the inequality is true and not true, you have likely solved the inequality correctly.

Tip

How do I know when I should add or subtract to solve an inequality?

As a general guideline, you should do the opposite of what is done in the inequality. When a number is subtracted from a variable in an inequality, such as $a - 5 > 11$, you can add the subtracted number to both sides of the inequality to cancel out the number and isolate the variable. If a number is added to a variable, such as $a + 5 > 11$, you can subtract the added number from both sides of the inequality to cancel out the number and isolate the variable.

Solve for the variable in each of the following inequalities. You can check your answers on page 256.

1) $x + 2 > 10$

2) $2x + 4 \geq x - 3$

3) $3 < x - 2$

4) $4 + x \leq 6$

5) $2 \geq x + 5$

6) $x + 2 < 8$

Solving Inequalities Example 2

$$6 \geq 14 + x$$
$$6 - 14 \geq 14 + x - 14$$
$$-8 \geq x$$
$$x \leq -8$$

Check your answer
Let $x = -8$

$6 \geq 14 + x$

$6 \geq 14 + (-8)$

$6 \geq 6$ *True!*

Check your answer
Let $x = -7$

$6 \geq 14 + x$

$6 \geq 14 + (-7)$

$6 \geq 7$ *Not true!*

Solving Inequalities Example 3

$$5x - 7 \leq 4x + 2$$
$$5x - 7 + 7 \leq 4x + 2 + 7$$
$$5x \leq 4x + 9$$
$$5x - 4x \leq 4x + 9 - 4x$$
$$x \leq 9$$

Check your answer
Let $x = 9$

$5x - 7 \leq 4x + 2$

$5(9) - 7 \leq 4(9) + 2$

$45 - 7 \leq 36 + 2$

$38 \leq 38$ *True!*

Check your answer
Let $x = 10$

$5x - 7 \leq 4x + 2$

$5(10) - 7 \leq 4(10) + 2$

$50 - 7 \leq 40 + 2$

$43 \leq 42$ *Not true!*

---

1 In this example, to place the variable by itself on one side of the inequality, subtract 14 from both sides of the inequality.

• If the variable ends up on the right side of an inequality, you can reverse the left and right sides of the inequality. You will also need to change the direction of the inequality sign. For example, $\geq$ changes to $\leq$.

2 To check your answer, choose two numbers that are closest to your answer—one number that should make the inequality true and another number that should make the inequality not true. Place the numbers into the original inequality and solve the problem.

• If the inequality is true and not true, you have likely solved the inequality correctly.

1 In this example, to place all the numbers on the right side of the inequality, add 7 to both sides of the inequality. You can then subtract $4x$ from both sides of the inequality to move all the variables to the left side of the inequality.

2 To check your answer, choose two numbers that are closest to your answer—one number that should make the inequality true and another number that should make the inequality not true. Place the numbers into the original inequality and solve the problem.

• If the inequality is true and not true, you have likely solved the inequality correctly.

Solve an Inequality with Multiplication and Division

You can use multiplication and division to remove or "cancel out" numbers and isolate the variable to one side of an inequality, such as $5x > 30$. This is called "solving for the variable."

When working with inequalities, you must perform the same operation on both sides of the inequality for the inequality to remain true. When a variable is divided by a number in an inequality, such as $\frac{x}{5} < 10$, you can multiply the same number on both sides of the inequality to cancel the number out and isolate the variable. If a variable is multiplied

by a number, such as $5x < 20$, you can divide both sides of the inequality by the same number to isolate the variable.

To check your answer, place two numbers closest to your answer—one that should make the inequality true and one that should make it false—into the inequality and solve the problem. For example, if $x > 5$, check the numbers 6 and 5. If 6 makes the inequality true and 5 makes the inequality false, you have likely solved the problem correctly.

How to Solve an Inequality

$$\frac{x}{3} \leq 5$$

$$\frac{x}{3} \times 3 \leq 5 \times 3$$

$$x \leq 15$$

$$6x > 30$$

$$\frac{6x}{6} > \frac{30}{6}$$

$$x > 5$$

Solving Inequalities Example 1

$$4x < 36$$

$$\frac{4x}{4} < \frac{36}{4}$$

$$x < 9$$

Check your answer
Let x = 8

$4x < 36$

$4(8) < 36$

$32 < 36$ *True!*

Check your answer
Let x = 9

$4x < 36$

$4(9) < 36$

$36 < 36$ *Not true!*

- When solving an inequality, your goal is to place the variable by itself on one side of the inequality, usually the left side, and place all the numbers on the other side of the inequality, usually the right side. This allows you to determine the value of the variable.

- To place a variable by itself on one side of an inequality, you can multiply or divide by the same number on both sides of the inequality and the inequality will remain true.

1 In this example, to place the variable by itself on the left side of the inequality, divide both sides of the inequality by the number in front of the variable, called the coefficient. This will leave the x variable by itself on the left side of the inequality because $4 \div 4$ equals 1.

2 To check your answer, choose two numbers that are closest to your answer—one number that should make the inequality true and another number that should make the inequality not true. Place the numbers into the original inequality and solve the problem.

- If the inequality is true and not true, you have likely solved the inequality correctly.

Important! When you multiply or divide by a negative number on both sides of an inequality, you must reverse the direction of the inequality symbol. For example, when working with the inequality $-5x > 30$, you divide both sides by -5 to isolate the x variable. When performing division, you reverse the $>$ symbol so that $x < -6$.

$$-5x > 30$$

$$\frac{-5x}{-5} < \frac{30}{-5}$$

$$x < -6$$

Practice
$a+b=c$

Solve for the variable in each of the following inequalities. You can check your answers on page 256.

1) $2x > 10$

2) $-5x \leq 25$

3) $4x \geq 7$

4) $\frac{-x}{3} < 10$

5) $\frac{2x}{5} > 4$

6) $\frac{3x}{4} \geq \frac{2}{3}$

Solving Inequalities Example 2

$$\frac{x}{2} \leq 16$$

$$\frac{x}{2} \times 2 \leq 16 \times 2$$

$$x \leq 32$$

Check your answer
Let $x = 32$

$$\frac{x}{2} \leq 16$$

$$\frac{32}{2} \leq 16$$

$$16 \leq 16 \text{ True!}$$

Check your answer
Let $x = 33$

$$\frac{x}{2} \leq 16$$

$$\frac{33}{2} \leq 16$$

$$16.5 \leq 16 \text{ Not true!}$$

1 In this example, you must multiply both sides of the inequality by the number divided into the variable to place the variable by itself on the left side of the inequality.

2 To check your answer, choose two numbers that are closest to your answer—one number that should make the inequality true and another number that should make the inequality not true. Place the numbers into the original inequality and solve the problem.

• If the inequality is true and not true, you have likely solved the inequality correctly.

Solving Inequalities Example 3

$$\frac{2}{3}x \geq 6$$

$$\frac{2}{3}x \times \frac{3}{2} \geq 6 \times \frac{3}{2}$$

$$x \geq \frac{18}{2}$$

$$x \geq 9$$

Check your answer
Let $x = 9$

$$\frac{2}{3}x \geq 6$$

$$\frac{2}{3}(9) \geq 6$$

$$\frac{18}{3} \geq 6$$

$$6 \geq 6 \text{ True!}$$

Check your answer
Let $x = 8$

$$\frac{2}{3}x \geq 6$$

$$\frac{2}{3}(8) \geq 6$$

$$\frac{16}{3} \geq 6$$

$$5.333... \geq 6 \text{ Not true!}$$

1 In this example, to place the variable by itself on the left side of the inequality, multiply both sides of the inequality by the reciprocal of the fraction in front of the variable.

Note: To determine the reciprocal of a fraction, switch the top and bottom numbers in the fraction. For example, $\frac{2}{3}$ becomes $\frac{3}{2}$. When you multiply a fraction by its reciprocal, the result is 1.

2 To check your answer, choose two numbers that are closest to your answer—one number that should make the inequality true and another number that should make the inequality not true. Place the numbers into the original inequality and solve the problem.

• If the inequality is true and not true, you have likely solved the inequality correctly.

Solve an Inequality in Several Steps

When solving complicated inequalities, your goal is to isolate the variable to one side of the inequality—just like when you are working out large equations. In the case of an inequality, however, the value of the variable will be a range of numbers.

There are three steps to follow when solving complex inequalities. First, simplify the inequality as much as you can on both sides of the inequality sign. Then use addition and subtraction on both sides of the inequality to isolate the variable on one side of the inequality. Next, multiply and

divide on both sides of the inequality to determine the value of the variable. Keep in mind that to keep the inequality true, you must reverse the direction of the inequality sign if you multiply or divide by a negative number.

Finally, check your answer. For the best results, place two numbers closest to your answer—one that should make the inequality true and one that should make it false—into the inequality and solve the problem. For example, if $x < 5$, check the numbers 4 and 5. If 4 makes the inequality true and 5 makes the inequality false, you have likely solved the problem correctly.

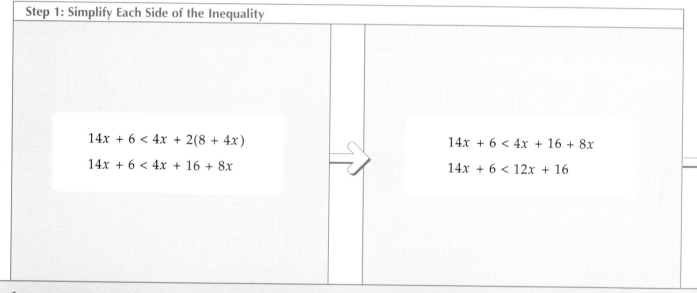

Step 1: Simplify Each Side of the Inequality

$$14x + 6 < 4x + 2(8 + 4x)$$
$$14x + 6 < 4x + 16 + 8x$$

$$14x + 6 < 4x + 16 + 8x$$
$$14x + 6 < 12x + 16$$

1 If an inequality contains parentheses ( ), use the distributive property on each side of the inequality to remove the parentheses that surround numbers and variables.

- In this example, $2(8 + 4x)$ works out to $16 + 8x$.

Note: The distributive property allows you to remove a set of parentheses by multiplying each number or variable inside the parentheses by a number or variable directly beside the parentheses. For more information on the distributive property, see page 30.

2 Add, subtract, multiply and divide any numbers or variables that can be combined on each side of the inequality.

- In this example, $4x + 8x$ equals $12x$.

Note: For information on adding and subtracting variables, see page 36. For information on multiplying and dividing variables, see page 38.

110

Tip

What should I do if I end up with a negative sign in front of a variable?

If you end up with a negative sign (–) in front of a variable, such as $-y > 14$, you will have to make the variable positive to finish solving the inequality. Fortunately, this is a simple task—just multiply both sides of the inequality by –1 and change the direction of the inequality symbol. In the case of $-y > 14$, you would finish with a solution of $y < -14$. Using this method, the inequality remains true while the variable switches from a negative to a positive value.

Practice

Solve for the variable in each of the following inequalities. You can check your answers on page 256.

1) $2(x - 5) < 8$

2) $-3(x + 1) \geq 6$

3) $x + 4 \leq 2x + x + 5$

4) $x - 3x + 9 > 4 + 3x$

5) $2(2x + 3) > -1(x + 4)$

6) $-4x \leq 20$

Step 2: Add and Subtract

$$14x + 6 < 12x + 16$$

$$14x + 6 - 6 < 12x + 16 - 6$$

$$14x < 12x + 10$$

$$14x - 12x < 12x + 10 - 12x$$

$$2x < 10$$

Step 3: Multiply and Divide

$$2x < 10$$

$$\frac{2x}{2} < \frac{10}{2}$$

$$x < 5$$

---

3 Determine which numbers and variables you need to add or subtract to place the variable by itself on one side of the inequality, usually the left side.

4 Add or subtract the numbers and/or variables you determined in step 3 on both sides of the inequality.

- In this example, subtract 6 from both sides of the inequality. You can then subtract $12x$ from both sides of the inequality.

  Note: For information on adding and subtracting numbers and variables in inequalities, see page 106.

5 Determine which numbers you need to multiply or divide by to place the variable by itself on one side of the inequality, usually the left side.

6 Multiply or divide by the numbers you determined in step 5 on both sides of the inequality.

- In this example, divide both sides of the inequality by 2. The variable x is left by itself on the left side of the inequality because $2 \div 2$ equals 1.

  Note: For information on multiplying and dividing numbers in inequalities, see page 108.

Solve a Compound Inequality

A compound inequality is like a variable sandwich—it features a variable surrounded by an expression on each side, such as $-12 < x < 10$. Compound inequalities contain three parts, unlike the usual two parts you find in equations and simple inequalities. The strategy for dealing with these three-part problems is actually quite simple and should be familiar to you. You must first simplify the inequality. Remove any parentheses by using the distributive property (see page 30) and then combine similar terms. Next, you isolate the

variable to the center of the problem by adding, subtracting, multiplying and dividing the same numbers and/or variables in all three parts of the inequality.

Once you have your inequality sandwich arranged with the variable in the middle, you should always check your main ingredient—the variable. Take the numbers determined by your solution for the variable, plug them into the original inequality and see if they work out.

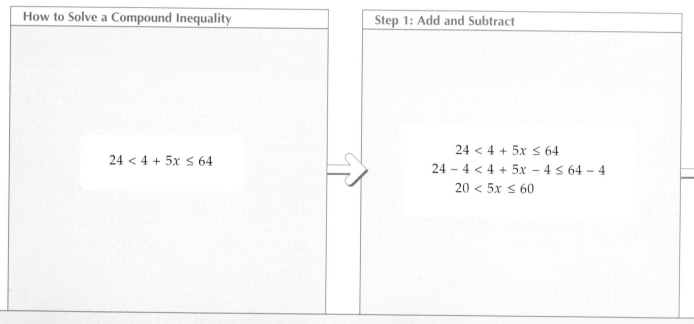

How to Solve a Compound Inequality	Step 1: Add and Subtract
$24 < 4 + 5x \leq 64$	$24 < 4 + 5x \leq 64$ $24 - 4 < 4 + 5x - 4 \leq 64 - 4$ $20 < 5x \leq 60$

- When solving a compound inequality, your goal is to place the variable by itself in the middle of the inequality.

- To place a variable by itself in the middle of an inequality, you can add, subtract, multiply or divide by the same number or variable in all three parts of the inequality and the inequality will remain true.

1 Determine which numbers and variables you need to add or subtract to place the variable by itself in the middle of the inequality.

2 Add and/or subtract the same numbers and/or variables you determined in step 1 in all three parts of the inequality.

- In this example, subtract 4 from all three parts of the inequality. By subtracting 4 from the middle of the inequality, you effectively cancel it out.

Note: For more information on adding and subtracting numbers and variables in inequalities, see page 106.

Important! When you multiply or divide all three parts of a compound inequality by a negative number, you must reverse the direction of the inequality symbols. For example, when working with the inequality $6 < -3x < 18$, when you divide all three parts of the inequality by -3 to isolate the x variable, make sure you reverse the $<$ symbols.

$$6 < -3x < 18$$

$$\frac{6}{-3} > \frac{-3x}{-3} > \frac{18}{-3}$$

$$-2 > x > -6$$

Practice Solve the following inequalities. You can check your answers on page 256.

1) $1 - 3 < x < 10 - 5$

2) $4 > 2x > 18$

3) $-6 \leq -3x \leq 12$

4) $0 \geq 2x - 4 \geq 8$

5) $-1 < 5 - x \leq 9$

6) $-2 \geq 6 + 2x > 4$

Step 2: Multiply and Divide

$$20 < 5x \leq 60$$

$$\frac{20}{5} < \frac{5x}{5} < \frac{60}{5}$$

$$4 < x \leq 12$$

Step 3: Check Your Answer

$$4 < x \leq 12$$

Check your answer
Let $x = 5$

$24 < 4 + 5x \leq 64$

$24 < 4 + 5(5) \leq 64$

$24 < 4 + 25 \leq 64$

$24 < 29 \leq 64$ *True!*

Check your answer
Let $x = 4$

$24 < 4 + 5x \leq 64$

$24 < 4 + 5(4) \leq 64$

$24 < 4 + 20 \leq 64$

$24 < 24 \leq 64$ *Not true!*

---

3 Determine which numbers you need to multiply or divide by to place the variable by itself in the middle of the inequality.

4 Multiply and/or divide all three parts of the inequality by the same numbers you determined in step 3.

- In this example, divide all three parts of the inequality by 5. The variable x remains by itself in the middle of the inequality because $5 \div 5$ equals 1.

  Note: For more information on multiplying and dividing numbers in inequalities, see page 108.

5 To check your answer, choose two numbers that are closest to the far left number in your answer—one number that should make the inequality true and another number that should make the inequality not true. Place the numbers into the original inequality and solve the problems. For example, since $4 < x$, check the numbers 5 and 4.

6 Repeat step 5 for two numbers that are closest to the far right number in your answer. For example, since $x \leq 12$, check the numbers 12 and 13.

- If the inequality is true and not true in both cases, you have likely solved the inequality correctly.

Solve an Inequality with Absolute Values

Solving inequalities that contain absolute values is complex because it combines techniques for dealing with absolute values with those used to solve inequalities. When solving inequalities with absolute values, remember that the absolute value of any non-zero number is the positive value of that number or variable, whether the number or variable is positive or negative. This means that $|a|$ can equal either a or $-a$.

The approach to inequalities containing absolute values varies depending on whether the simplified inequalities contain absolute values that are greater than or less than the other side of the

inequality. You must first simplify the inequality you want to work with and isolate the absolute value expression on one side of the inequality. If the simplified inequality contains an absolute value that is greater than or equal to the other side of the inequality, you split the inequality into two different inequalities. Then you can solve each of the two new inequalities and determine the possible answers for the variable.

You should check your work by plugging numbers from the solutions into the original inequality and working out the problems to verify your answers.

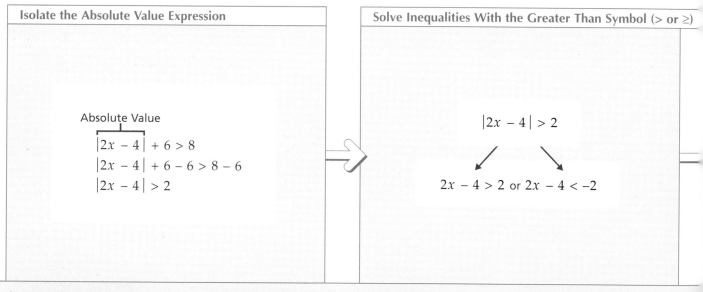

Isolate the Absolute Value Expression

Absolute Value

$|2x - 4| + 6 > 8$

$|2x - 4| + 6 - 6 > 8 - 6$

$|2x - 4| > 2$

Solve Inequalities With the Greater Than Symbol (> or ≥)

$|2x - 4| > 2$

$2x - 4 > 2$ or $2x - 4 < -2$

1 Determine which numbers and variables you need to add, subtract, multiply or divide to place the absolute value expression by itself on the left side of the inequality.

2 Add, subtract, multiply or divide both sides of the inequality by the numbers and/or variables you determined in step 1.

• In this example, subtract 6 from both sides of the inequality.

Note: For information on adding, subtracting, multiplying and dividing numbers and variables in inequalities, see pages 106 to 109.

• You now need to create two new inequalities from the original inequality.

Note: If the inequality contains the less than symbol (< or ≤), see page 116.

3 Create two new inequalities that look like the original inequality, but without the absolute value symbols (||).

4 For the second inequality, place a negative sign (–) in front of the value that the absolute value expression is greater than (> or ≥). Then change the direction of the inequality symbol from > to < or from ≥ to ≤.

Tip

When I create two new inequalities to solve an inequality with absolute values, why do I place "or" between the two inequalities?

It is standard form to write "or" between the two new inequalities you create because a value that you try when checking your answers only has to be true in one of the new inequalities—not both—to be a solution to the original inequality.

$$|2x - 4| > 2$$

$$2x - 4 > 2 \text{ or } 2x - 4 < -2$$

Practice

$a + b = c$

Solve the following inequalities containing absolute values with greater than symbols. You can check your answers on page 257.

1) $|x| > 2$

2) $|x| > -1$

3) $|x| > 0$

4) $|2x| \geq 8$

5) $|3x - 2| > 7$

6) $|4 - 2x| \geq 8$

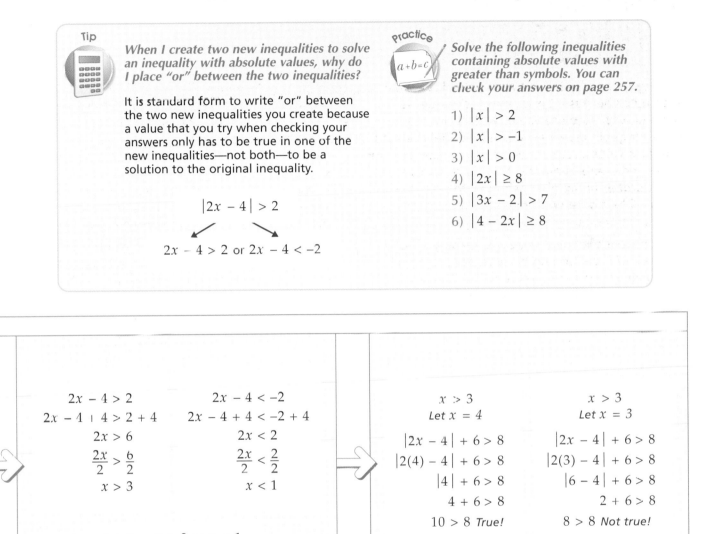

$$2x - 4 > 2 \qquad\qquad 2x - 4 < -2$$
$$2x - 4 + 4 > 2 + 4 \qquad 2x - 4 + 4 < -2 + 4$$
$$2x > 6 \qquad\qquad 2x < 2$$
$$\frac{2x}{2} > \frac{6}{2} \qquad\qquad \frac{2x}{2} < \frac{2}{2}$$
$$x > 3 \qquad\qquad x < 1$$

Solution: $x > 3$ or $x < 1$

$$x > 3$$
$$\text{Let } x = 4$$
$$|2x - 4| + 6 > 8$$
$$|2(4) - 4| + 6 > 8$$
$$|4| + 6 > 8$$
$$4 + 6 > 8$$
$$10 > 8 \text{ True!}$$

$$x > 3$$
$$\text{Let } x = 3$$
$$|2x - 4| + 6 > 8$$
$$|2(3) - 4| + 6 > 8$$
$$|6 - 4| + 6 > 8$$
$$2 + 6 > 8$$
$$8 > 8 \text{ Not true!}$$

- You now need to solve the two new inequalities separately in order to determine the answers for the original inequality.

5 In both inequalities, determine which numbers and variables you need to add, subtract, multiply or divide to place the variable by itself on one side of the inequality.

6 Add, subtract, multiply or divide both sides of the inequality by the numbers and/or variables you determined in step 5.

- In both examples, add 4 to both sides of the inequality. Then divide by 2 on both sides of the inequality.

7 To check your answer, choose two numbers that are closest to your first answer—one number that should make the inequality true and another number that should make the inequality not true. Place the numbers into the original inequality and solve the problem. For example, if $x > 3$, you would select 4 and 3.

8 Repeat step 7 with two numbers from your second answer. For example, if $x < 1$, you would select 0 and 1.

- If the inequality is true and not true in steps 7 and 8, you have likely solved the inequality correctly.

CONTINUED

Solve an Inequality with Absolute Values *continued*

As mentioned earlier, solving inequalities that contain absolute values is complicated due to the fact that you must combine techniques for dealing with absolute values and the rules used to solve inequalities. Remember that the approach varies, depending on whether the simplified inequalities contain absolute values that are greater than or less than the other side of the inequality.

First, simplify an inequality and isolate the absolute value expression on one side of the inequality. If the inequality contains an absolute value that is less than or equal to the other side of the inequality, you create a compound inequality by removing the absolute value symbols and writing the value from the right side of the inequality on the left side of the inequality with a negative sign beside it. Then place the same inequality symbol between the new value and what used to be the absolute value expression. For example, from $|x + 5| \leq 17$ you would create $-17 \leq x + 5 \leq 17$. Next, you solve for the variable and, finally, check your answer.

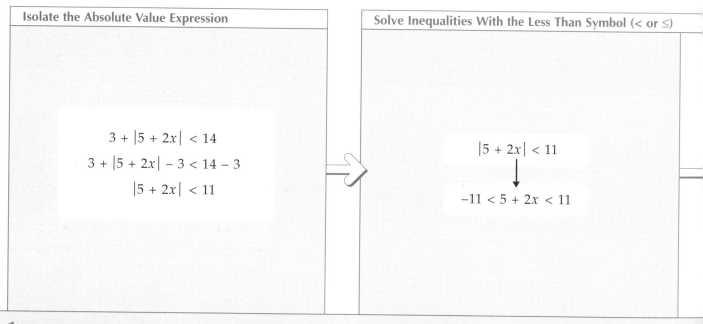

Isolate the Absolute Value Expression

$$3 + |5 + 2x| < 14$$
$$3 + |5 + 2x| - 3 < 14 - 3$$
$$|5 + 2x| < 11$$

Solve Inequalities With the Less Than Symbol (< or ≤)

$$|5 + 2x| < 11$$
$$-11 < 5 + 2x < 11$$

1 Determine which numbers and variables you need to add, subtract, multiply or divide to place the absolute value expression by itself on the left side of the inequality.

2 Add, subtract, multiply or divide both sides of the inequality by the numbers and/or variables you determined in step 1.

• In this example, subtract 3 from both sides of the inequality.

Note: For information on adding, subtracting, multiplying and dividing numbers and variables in inequalities, see pages 106 to 109.

• You now need to create a compound inequality from the original inequality.

3 Create a new inequality from the original inequality, but without the absolute value symbols (||).

4 Write the same number from the right side of the inequality on the left side, but place a negative sign (–) in front of the number. Then write the same inequality symbol (< or ≤) to the left of the absolute value expression.

Tip

Is there a solution to every inequality with absolute values?

No. As you work with inequalities that contain absolute values, you will sometimes come across inequalities that have no solution. One dead give-away occurs when the absolute value is less than or equal to a negative number, as in $|4x + 2| < -10$. Don't even bother working out these inequalities. You will also find inequalities that have only one solution. An absolute value expression that is less than or equal to zero, such as $|x| \leq 0$, has only one solution, which is $x = 0$.

Practice

$a + b = c$

Solve the following inequalities containing absolute values with less than symbols. You can check your answers on page 257.

1) $|x| < 5$

2) $|x| < -2$

3) $|x| \leq 0$

4) $|7x| \leq 21$

5) $|x - 4| < 5$

6) $|3x - 2| \leq 4$

$$-11 < 5 + 2x < 11$$
$$-11 - 5 < 5 + 2x - 5 < 11 - 5$$
$$16 < 2x < 6$$
$$\frac{-16}{2} < \frac{2x}{2} < \frac{6}{2}$$
$$-8 < x < 3$$

$-8 < x < 3$
Let x = -7

$3 + |5 + 2x| < 14$
$3 + |5 + 2(-7)| < 14$
$3 + |5 + (-14)| < 14$
$3 + |(-9)| < 14$
$3 + 9 < 14$
$12 < 14$ *True!*

$-8 < x < 3$
Let x = -8

$3 + |5 + 2x| < 14$
$3 + |5 + 2(-8)| < 14$
$3 + |5 + (-16)| < 14$
$3 + |-11| < 14$
$3 + 11 < 14$
$14 < 14$ *Not true!*

---

5 Determine which numbers and variables you need to add, subtract, multiply or divide to place the variable by itself in the middle of the inequality.

6 Add, subtract, multiply or divide the numbers and/or variables you determined in step **5** in all three parts of the inequality.

- In this example, subtract 5 from all three parts of the inequality. Then divide all three parts of the inequality by 2.

Note: For more information on solving compound inequalities, see page 112.

7 To check your answer, choose two numbers that are closest to the far left number in your answer—one number that should make the inequality true and another number that should make the inequality not true. Place the numbers into the original inequality and solve the problems. For example, since $-8 < x$, check the numbers -7 and -8.

8 Repeat step **7** for two numbers that are closest to the far right number in your answer. For example, since $x < 3$, check the numbers 2 and 3.

- If the inequality is true and not true in both cases, you have likely solved the inequality correctly.

Graph an Inequality with One Variable

Some people learn better visually. If you are one of those people, you are going to love number lines. Number lines can be used to graph inequalities, visually indicating all the possible solutions.

Basic inequalities, such as $x > 5$, have an infinite number of solutions. When graphing basic inequalities, you use a line that begins with a circle and ends with an arrow which indicates that the solution does not have an upper or a lower boundary. A compound inequality, such as $-1 \leq x < 3$, has a limited range of possible values. When graphing compound inequalities, you use a line with circles on both ends to show the upper and lower boundaries for the range of values.

You can adjust the numbers shown on a number line to best suit the inequality you are graphing. For example, to graph the inequality $x > 25$, you can draw a number line that is numbered from 20 to 30.

Graphing Basic Inequalities

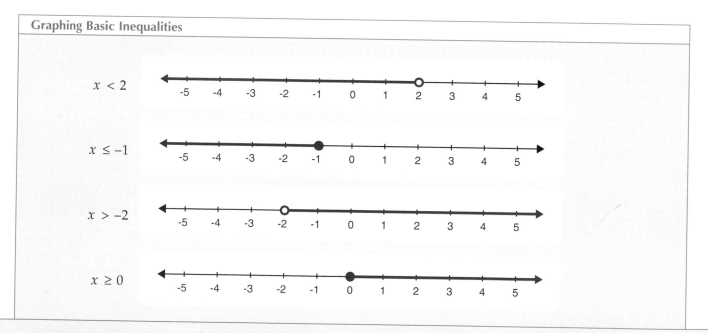

- You can use a number line to graph basic inequalities, such as $x < 2$.

1 To graph a basic inequality, locate the number in the inequality on the number line. For example, in the inequality $x < 2$, locate 2 on the number line.

2 If the inequality symbol is $<$ or $>$, draw a hollow circle (o) at the location of the number on the number line. If the inequality symbol is $\leq$ or $\geq$, draw a filled-in circle (•) at the location of the number on the number line.

- A hollow circle (o) on the number line indicates the number is not a solution. A filled-in circle (•) indicates the number is a solution.

3 If the inequality symbol is $<$ or $\leq$, draw an arrow on the number line from the circle to the left end of the number line. If the inequality symbol is $>$ or $\geq$, draw an arrow on the number line from the circle to the right end of the number line.

- The line you draw indicates all the possible solutions to the inequality.

Practice

$a+b=c$

Graph the following inequalities. You can check your answers on page 257.

1) $x > -4$

2) $x \geq -3$

3) $x \leq 1$

4) $-2 < x < 1$

5) $1 \leq x \leq 5$

6) $-3 < x \leq 0$

Use inequalities to describe the following graphs. You can check your answers on page 257.

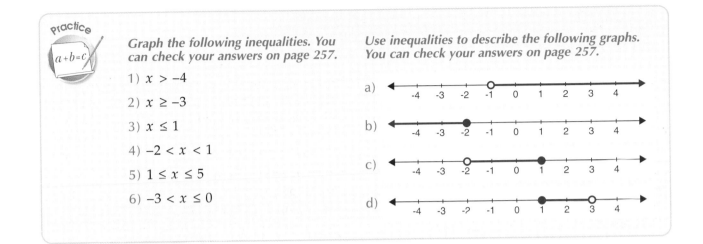

Graphing Compound Inequalities

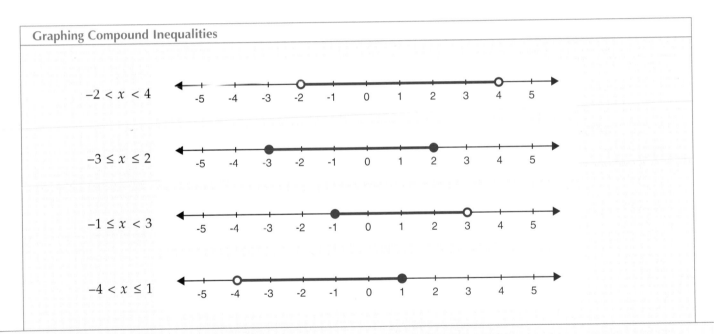

- You can use a number line to graph compound inequalities, such as $-2 < x < 4$.

1 To graph a compound inequality, locate the first number in the inequality on the number line. For example, in the inequality $-2 < x < 4$, locate -2 on the number line.

2 If the inequality symbol is < or >, draw a hollow circle (○) at the location of the number on the number line. If the inequality symbol is ≤ or ≥, draw a filled-in circle (●) at the location of the number on the number line.

3 Repeat steps 1 and 2 for the second number in the inequality. For example, in the inequality $-2 < x < 4$, mark 4 on the number line.

4 Draw a line on the number line from one circle to the other circle.

- The line you draw indicates all the possible solutions to the inequality.

Note: A compound inequality such as $-2 < x < 4$ can also be written as $x > -2$ and $x < 4$.

119

Graph an Inequality with Two Variables

A linear inequality contains one or two variables and appears as a line and a shaded area in a coordinate plane. To graph a linear inequality, pretend that the inequality symbol is an equals sign. If the inequality contains a ≤ or a ≥ symbol, draw a solid line. If the inequality contains a < or > symbol, however, you need to draw a dashed line, indicating that the coordinates along the line are not solutions for the inequality.

Next you must shade the side of the line that contains all of the points that are solutions to the

inequality. To determine which side of the line should be shaded, select a random coordinate pair. Most people pick the origin point (0,0) because working with 0 is so easy. Whichever point you choose, plug the x and/or y values into the inequality and work out the problem. If the inequality is true, the point you chose solves the inequality and you shade the side of the line that the point appears on. If the inequality is not true, shade the opposite side of the line.

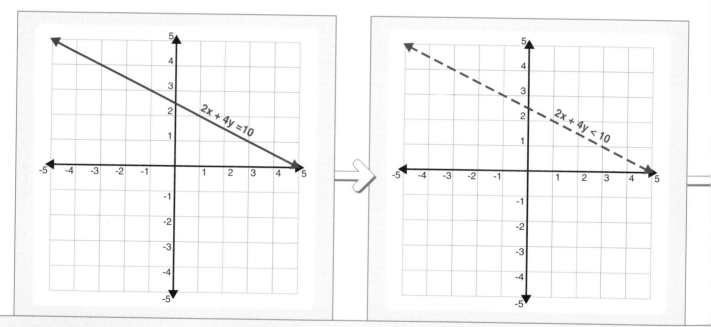

• You will use a coordinate plane to graph linear inequalities.

1 To graph a linear inequality, pretend the inequality symbol (<, ≤, > or ≥) is an equals sign (=) and graph the resulting line in a coordinate plane.

• In this example, for the inequality $2x + 4y < 10$, graph the line $2x + 4y = 10$.

Note: For information on graphing lines, see page 84.

2 If the inequality symbol is < or >, change the solid line to a dashed line to indicate that the points along the line are not solutions to the inequality.

• If the inequality symbol is ≤ or ≥, keep the solid line to indicate that the points along the line are solutions to the inequality.

Tip

What is the difference between the graph of a linear inequality and the graph of a linear equation?

Unlike the graph of a linear equation, which is just a solid line, the graph of a linear inequality consists of a solid or dashed line as well as a shaded area in the coordinate plane. All of the points within the shaded area represent solutions to the inequality. For more information on graphing linear equations, see page 84.

Practice

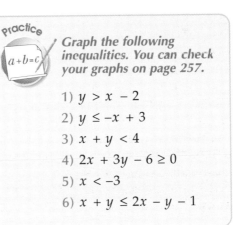

Graph the following inequalities. You can check your graphs on page 257.

1) $y > x - 2$

2) $y \leq -x + 3$

3) $x + y < 4$

4) $2x + 3y - 6 \geq 0$

5) $x < -3$

6) $x + y \leq 2x - y - 1$

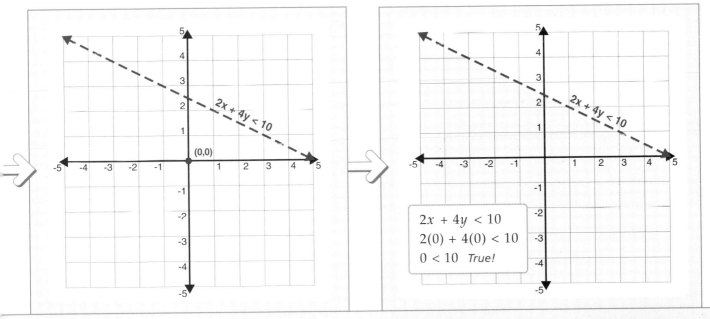

- The line splits the coordinate plane into two parts. The points on one side of the line will make the inequality true. The points on the other side of the line will not make the inequality true.

3 To determine which side of the line contains the points that will make the inequality true, select any point on either side of the line.

Note: Selecting the point (0,0) will make the problem easy to solve.

4 Place the numbers for the point into the inequality and solve the problem.

5 If the inequality is true, lightly shade the part of the coordinate plane that contains the point. All the points in this side of the coordinate plane are solutions to the inequality.

- If the inequality is not true, lightly shade the other part of the coordinate plane that does not contain the point. All the points in the shaded side of the coordinate plane are solutions to the inequality.

Solving and Graphing Inequalities

Question 1. Solve each of the following inequalities.

a) $x + 2 > 5$

b) $x - 4 < 0$

c) $2 - y \geq 7$

d) $2 + x \leq 2x - 10$

e) $4x < 0$

f) $-x > 5$

g) $-3y \geq 9$

h) $4x - 1 < 2x + 5$

i) $2 - x > 2 + 3x$

Question 2. Solve each of the following compound inequalities.

a) $1 < x + 2 \leq 5$

b) $3 < x - 4 < 7$

c) $1 \leq -x \leq 4$

d) $-2 \leq 2 - 3x < 0$

e) $x + 1 < 3x - 2 \leq 4 + x$

f) $5 - x \leq x \leq -x + 7$

g) $2x + 1 < x < 5 + 2x$

Question 3. Solve each of the following inequalities.

a) $|x| < 2$

b) $|x| > 3$

c) $|-2x| < 10$

d) $|x - 1| \leq 4$

e) $|2x + 3| \geq 1$

f) $|5x - 10| < 8$

g) $|3 - x| < 0$

Question 4. Graph the following inequalities on a number line.

a) $x > 3$

b) $x < 1$

c) $x \geq 10$

d) $x \leq -3$

e) $1 < x \leq 3$

f) $-2 \leq x \leq 6$

g) $x < 2$ or $x > 7$

h) $x \leq -1$ or $x \geq 0$

i) $x < 0$ or $x \geq 5$

Question 5. Graph the following inequalities on a number line.

a) $2x + 4 < 10$

b) $2 - x \geq 5$

c) $|x| < 5$

d) $|-x| < 2$

e) $|x - 1| < 3$

f) $|2x + 1| \geq 4$

g) $|5 - x| \leq 3$

Question 6. Graph the following inequalities in a coordinate plane.

a) $y > x + 1$

b) $y \geq 2 - x$

c) $y \leq 2x + 5$

d) $y < \dfrac{x}{2}$

e) $x + y > 0$

f) $x - y < 1$

g) $2x + 3y \geq 4$

h) $y + 2x - 3 \leq 0$

*You can check your answers
on pages 271-273.*

123

Chapter 6

Despite the science fiction connotation, a matrix in algebra is nothing more exciting than a bunch of numbers grouped together by brackets. There are a number of special rules that govern the way you must treat matrices. Chapter 6 explains what matrices are and how to work with them.

Working with Matrices

In this Chapter...

Introduction to Matrices

A matrix is a mathematical term for a group of values which are arranged in horizontal rows and vertical columns, usually enclosed in brackets. Since they are so nicely organized, matrices are an excellent way to present and work with large amounts of data. Spreadsheets are a form of matrix that you may be familiar with.

Each matrix is classified by its order, which is the number of horizontal rows and vertical columns of information that are contained within the matrix, with the number of rows always listed first. For example, a matrix with 2 rows and 3

columns is said to have an order of 2 x 3, read as "2 by 3." A matrix with an equal number of rows and columns is referred to as a square matrix.

Each value, called an element, in a matrix is named using a lowercase letter that matches the uppercase name of the matrix, followed by two small numbers below and to the right of the letter, such as a_{13}. The first small number indicates the row where the element is found, while the second small number indicates the column. For larger matrices, a comma is used to separate the numbers and avoid confusion.

About a Matrix

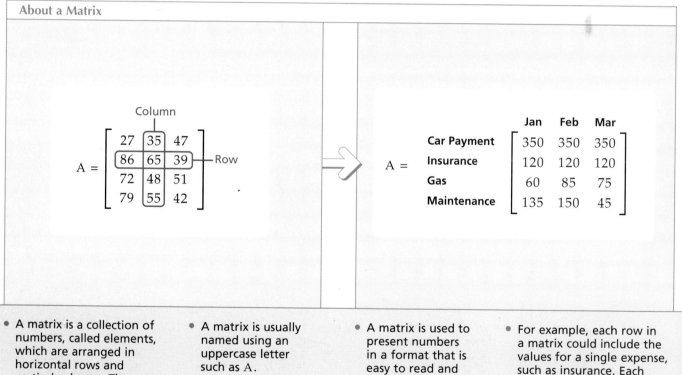

- A matrix is a collection of numbers, called elements, which are arranged in horizontal rows and vertical columns. The collection of numbers is surrounded by brackets.

- A matrix is usually named using an uppercase letter such as A.

  Note: Matrices is the term used to indicate more than one matrix.

- A matrix is used to present numbers in a format that is easy to read and understand.

- For example, each row in a matrix could include the values for a single expense, such as insurance. Each column in a matrix could include the values for a specific time period, such as one month.

Practice $a+b=c$

Determine the order of the following matrices. Then name each element in each matrix, such as $a_{11} = 7$. You can check your answers on page 258.

$$A = \begin{bmatrix} 7 & 5 \\ -3 & 8 \\ 2 & -6 \end{bmatrix} \quad B = \begin{bmatrix} 5 & 3 & -2 \\ -9 & 6 & 4 \end{bmatrix}$$

The Order of a Matrix

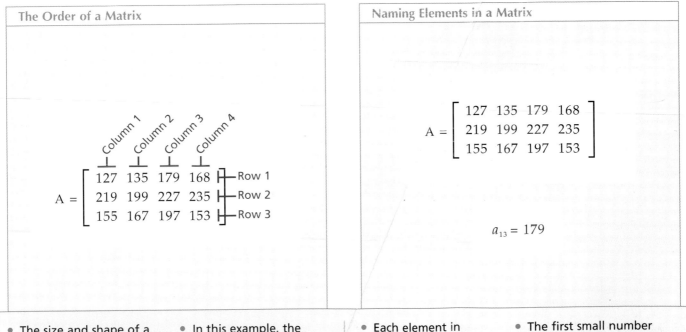

$$A = \begin{matrix} & \text{Column 1} & \text{Column 2} & \text{Column 3} & \text{Column 4} \\ \begin{bmatrix} 127 & 135 & 179 & 168 \\ 219 & 199 & 227 & 235 \\ 155 & 167 & 197 & 153 \end{bmatrix} & \text{Row 1} \\ & \text{Row 2} \\ & \text{Row 3} \end{matrix}$$

- The size and shape of a matrix is referred to as the order of the matrix.

- The order of a matrix indicates the number of rows and columns in the matrix and is written as two numbers separated by an x, such as 3 x 4. The number of rows is always given before the number of columns.

- In this example, the matrix has 3 rows and 4 columns. The order of the matrix is 3 x 4, which is read as 3 "by" 4.

  Note: When a matrix has the same number of rows and columns, the matrix is referred to as a square matrix.

Naming Elements in a Matrix

$$A = \begin{bmatrix} 127 & 135 & 179 & 168 \\ 219 & 199 & 227 & 235 \\ 155 & 167 & 197 & 153 \end{bmatrix}$$

$$a_{13} = 179$$

- Each element in a matrix is named using a lowercase letter that matches the name of the matrix, followed by two small numbers below and to the right of the letter, such as a_{13}.

- The first small number following the letter indicates the row where the element is found in the matrix. The second small number indicates the column where the element is found in the matrix.

- For example, the element named a_{13} is located in the first row and the third column in the matrix.

Add and Subtract Matrices

To add or subtract matrices, you must first check whether the matrices have the same order, which is the number of horizontal rows and vertical columns within a matrix. Only matrices of the same order can be added or subtracted.

To add matrices together, add the corresponding elements in each matrix and place the answers in a third matrix that has the same order as the matrices you are adding. For example, add the value in the top left position in the first matrix to the value in the top left position in the second matrix. Then place the answer in the top left position in a new matrix. Repeat these steps for each element in the matrices. You subtract matrices in the same fashion, subtracting the elements in one matrix from the corresponding elements in another.

There are many real life examples of why you might want to add or subtract matrices. For instance, say you run a clothing store and want to gauge the skills of your sales staff. You could create a matrix for each month, with rows representing each salesperson and columns representing sales in each category of clothing. You could then subtract the matrices to find each salesperson's monthly increase or decrease in sales for each clothing category.

Add Matrices

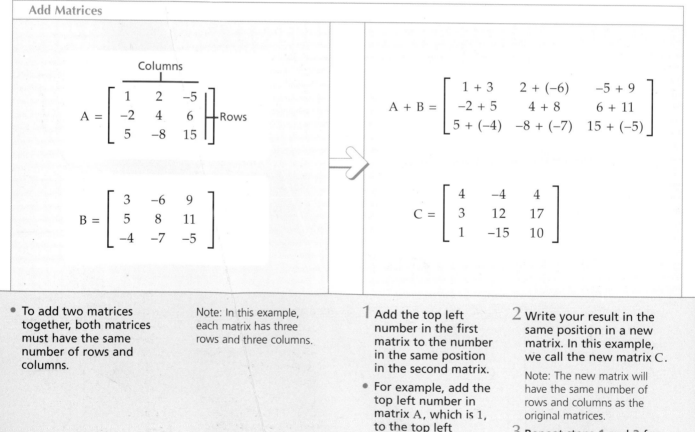

- To add two matrices together, both matrices must have the same number of rows and columns.

Note: In this example, each matrix has three rows and three columns.

1 Add the top left number in the first matrix to the number in the same position in the second matrix.

- For example, add the top left number in matrix A, which is 1, to the top left number in matrix B, which is 3.

2 Write your result in the same position in a new matrix. In this example, we call the new matrix C.

Note: The new matrix will have the same number of rows and columns as the original matrices.

3 Repeat steps 1 and 2 for each pair of corresponding numbers in the matrices.

Practice

$a+b=c$

1) Add matrix A and matrix B and write the results in a new matrix labelled C. You can check your answers on page 258.

2) Subtract matrix B from matrix A and write the results in a new matrix labelled D. You can check your answers on page 258.

$$A = \begin{bmatrix} 6 & -3 & -8 \\ -7 & 5 & 1 \\ 4 & 2 & -5 \end{bmatrix} \quad B = \begin{bmatrix} 5 & -9 & 1 \\ -3 & -10 & 4 \\ 6 & -8 & 2 \end{bmatrix}$$

Subtract Matrices

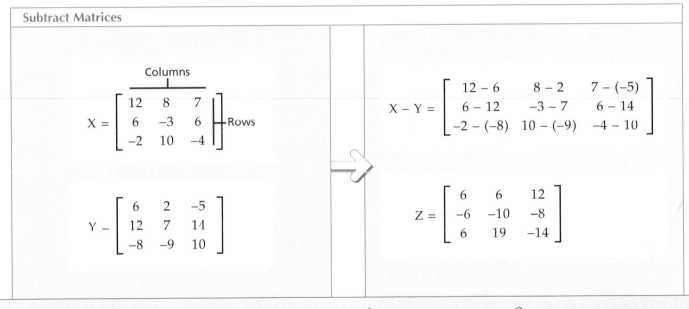

Columns

$$X = \begin{bmatrix} 12 & 8 & 7 \\ 6 & -3 & 6 \\ -2 & 10 & -4 \end{bmatrix} \text{Rows}$$

$$Y - \begin{bmatrix} 6 & 2 & -5 \\ 12 & 7 & 14 \\ -8 & -9 & 10 \end{bmatrix}$$

$$X - Y = \begin{bmatrix} 12-6 & 8-2 & 7-(-5) \\ 6-12 & -3-7 & 6-14 \\ -2-(-8) & 10-(-9) & -4-10 \end{bmatrix}$$

$$Z = \begin{bmatrix} 6 & 6 & 12 \\ -6 & -10 & -8 \\ 6 & 19 & -14 \end{bmatrix}$$

- To subtract matrices, both matrices must have the same number of rows and columns.

Note: In this example, each matrix has three rows and three columns.

1 Subtract the top left number in the second matrix from the number in the same position in the first matrix.

- For example, subtract the top left number in matrix Y, which is 6, from the top left number in matrix X, which is 12.

2 Write your result in the same position in a new matrix. In this example, we call the new matrix Z.

Note: The new matrix will have the same number of rows and columns as the original matrices.

3 Repeat steps 1 and 2 for each pair of corresponding numbers in the matrices.

Multiply a Matrix by a Scalar

A matrix can easily be doubled, tripled or multiplied by any value. The value you multiply the matrix by is called a scalar. The term scalar simply means a value that is not enclosed inside the matrix.

Fortunately, scalar multiplication is one of the easiest operations that you can perform on a matrix. You simply multiply each element, or value within the matrix, by the scalar, creating a new matrix with the same order, or dimensions,

as the original matrix. You place the results of the multiplication in the same positions in the new answer matrix as the elements in the original matrix.

You can multiply a matrix by both positive and negative scalars. When you are performing scalar multiplication with a negative value, make sure to change the signs in your results—positive numbers will become negative, while negative numbers will become positive.

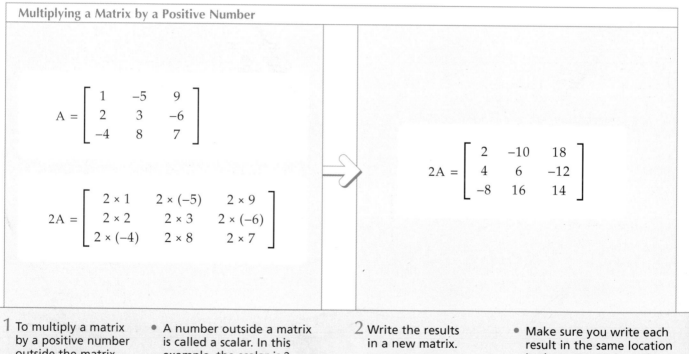

Multiplying a Matrix by a Positive Number

$$A = \begin{bmatrix} 1 & -5 & 9 \\ 2 & 3 & -6 \\ -4 & 8 & 7 \end{bmatrix}$$

$$2A = \begin{bmatrix} 2 \times 1 & 2 \times (-5) & 2 \times 9 \\ 2 \times 2 & 2 \times 3 & 2 \times (-6) \\ 2 \times (-4) & 2 \times 8 & 2 \times 7 \end{bmatrix}$$

$$2A = \begin{bmatrix} 2 & -10 & 18 \\ 4 & 6 & -12 \\ -8 & 16 & 14 \end{bmatrix}$$

1 To multiply a matrix by a positive number outside the matrix, multiply each number inside the matrix by the number outside the matrix.

- A number outside a matrix is called a scalar. In this example, the scalar is 2.

- In this example, since we are multiplying matrix A by the number 2, we call the new matrix 2A.

2 Write the results in a new matrix.

- The new matrix will have the same number of rows and columns as the original matrix.

- Make sure you write each result in the same location in the new matrix as the original location in the original matrix.

Note: When you multiply two numbers with different signs (+ or −), the result will always be a negative number. For example, 2 × (−4) equals −8.

Practice

$a + b = c$

1) *Multiply the following matrix by 3 and write your results in a new matrix labelled 3A. You can check your answers on page 258.*

2) *Multiply the following matrix by −5 and write the results in a new matrix labelled −5A. You can check your answers on page 258.*

$$A = \begin{bmatrix} -1 & 7 & -5 \\ 2 & 6 & 4 \\ 8 & -3 & -9 \end{bmatrix}$$

Multiplying a Matrix by a Negative Number

$$B = \begin{bmatrix} 6 & 8 & -3 \\ -9 & -2 & 1 \\ 5 & 4 & -7 \end{bmatrix}$$

$$-3B = \begin{bmatrix} -3 \times 6 & -3 \times 8 & -3 \times (-3) \\ -3 \times (9) & -3 \times (-2) & -3 \times 1 \\ -3 \times 5 & -3 \times 4 & 3 \times (-7) \end{bmatrix}$$

$$-3B = \begin{bmatrix} -18 & -24 & 9 \\ 27 & 6 & -3 \\ -15 & -12 & 21 \end{bmatrix}$$

1 To multiply a matrix by a negative number outside the matrix, multiply each number inside the matrix by the number outside the matrix.

- A number outside a matrix is called a scalar. In this example, the scalar is −3.

- In this example, since we are multiplying matrix B by the number −3, we call the new matrix −3B.

2 Write the results in a new matrix.

- The new matrix will have the same number of rows and columns as the original matrix.

- Make sure you write each result in the same location in the new matrix as the original location in the original matrix.

Note: When you multiply two numbers with the same sign (+ or −), the result will always be a positive number. For example, (−3) × (−9) equals 27.

131

Multiply Matrices

When multiplying matrices, the dimensions of the two matrices do not have to be the same, but the number of columns in the first matrix must equal the number of rows in the second matrix. The answer matrix will have the same number of rows as the first matrix and the same number of columns as the second matrix.

A couple of patterns help you easily determine which matrices can be multiplied together and the dimensions of the new answer matrix. For example, you can multiply a 2 x 4 matrix by a 4 x 6 matrix. In this case, notice that the middle numbers—the 4s—are the same. Additionally, the result of multiplying these two matrices will be a 2 x 6 matrix. Notice that the outside numbers of the dimensions of the original matrices—2 and 6—are the dimensions of the answer matrix.

For your answer, create a new matrix using the dimensions that you have already determined. Label the locations for each number, or element, within the matrix by row and column. For example, x_{12} would be the label for the element in the first row and second column of the answer matrix.

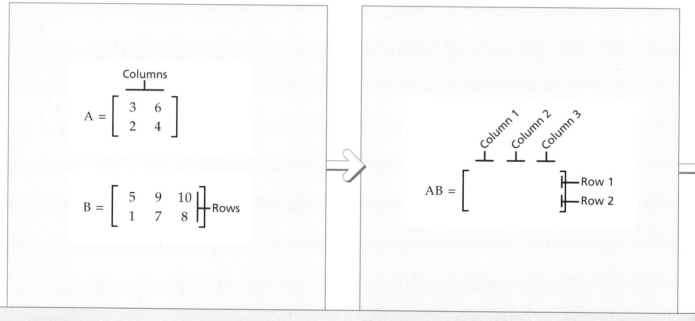

- To multiply two matrices together, the number of columns in the first matrix must equal the number of rows in the second matrix.

- In this example, matrix A has 2 columns and matrix B has 2 rows, so you can multiply the matrices.

1 To determine the size of the new matrix in which you will enter your multiplication results, use the number of rows in the first matrix and the number of columns in the second matrix.

- In this example, matrix A has 2 rows and matrix B has 3 columns, so the new matrix will have 2 rows and 3 columns.

2 Create a matrix of the size you determined in step 1.

- In this example, since we are multiplying matrix A by matrix B, we call the new matrix AB.

Tip

Can I multiply matrices in any order?

No. The order in which you multiply matrices is important since the order will often change the result. For example, if you have two square matrices (*A* and *B*), if you multiply matrix *A* by matrix *B*, and then multiply matrix *B* by matrix *A*, you will often end up with two different results. This is not the case when multiplying numbers. The order in which you multiply numbers will not change the result. For example, whether you multiply 2 × 5 or 5 × 2, you will end up with 10 in both cases.

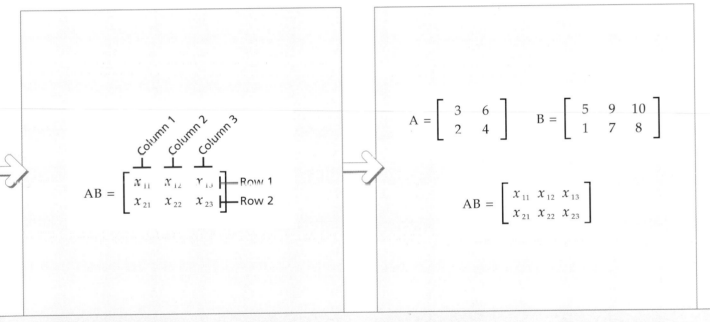

3 Name each position in the new matrix using a variable, such as x, followed by two small numbers, known as subscripts. The first subscript indicates the row where the variable is located in the new matrix. The second subscript indicates the column where the variable is located in the new matrix.

• For example, the position in the first row and first column in the new matrix is named x_{11}. The position in the first row and second column in the new matrix is named x_{12}, and so on.

• To find the value of one number in the new matrix, such as x_{11}, you will multiply all the numbers in an entire row in the first matrix by all the corresponding numbers in an entire column in the second matrix. You will add the results together to determine the value of the number.

• The first subscript beside each x indicates the row you will use in the first matrix. The second subscript indicates the column you will use in the second matrix.

• For example, x_{12} tells you to use the first row in the first matrix and the second column in the second matrix.

Multiply Matrices
continued

The matrix you created to contain the product of your two original matrices becomes very helpful when you start to multiply the elements.

The numbers in the label of each element in the answer matrix tell you which row in the first matrix and which column in the second matrix you must work with. For example, the label x_{21} tells you that you will be working with the second row of the first matrix and the first column of the second matrix. Now that you know which row and column you are working with, multiply the elements in the row with the corresponding elements in the column. Once you have all the products of the multiplications,

you add them all together and place the sum in the answer matrix in the same location.

Because this process requires so much looking back and forth between matrices, you may find it helpful to get your hands involved. With your left hand, point to the first element of the row in the first matrix and with your right hand, point to the first element in the column of the second matrix. Multiply those two numbers and then move your left finger across the row by one number and your right finger down the column by one number. Keep multiplying until you get to the end of each row and column.

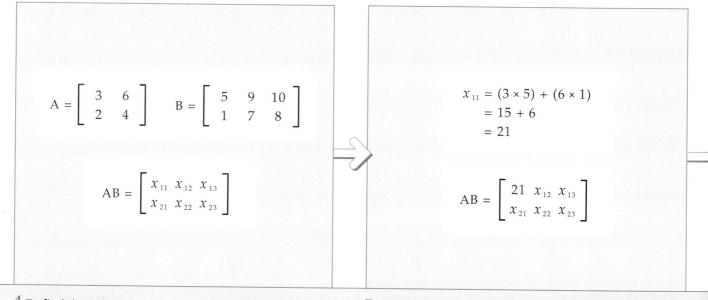

4 To find the value of one number in the new matrix, look at the subscripts beside an x in the new matrix to determine the row you will use in the first matrix and the column you will use in the second matrix.

• For example, x_{11} indicates the first row in the first matrix and the first column in the second matrix.

5 Using the row and column you determined in step 4, multiply each number in the row in the first matrix by the corresponding number in the column in the second matrix.

Note: Start by multiplying the first number in both the row and column. Then multiply the second number in both the row and column, and so on until you reach the end of the row and column.

6 When you reach the end of the row and column, add the results together.

7 Write your result in the corresponding position in the new matrix, replacing the variable x and the subscripts.

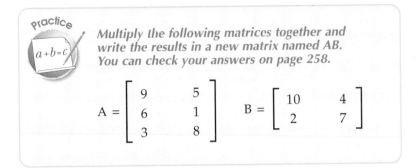

Multiply the following matrices together and write the results in a new matrix named AB. You can check your answers on page 258.

$$A = \begin{bmatrix} 9 & 5 \\ 6 & 1 \\ 3 & 8 \end{bmatrix} \quad B = \begin{bmatrix} 10 & 4 \\ 2 & 7 \end{bmatrix}$$

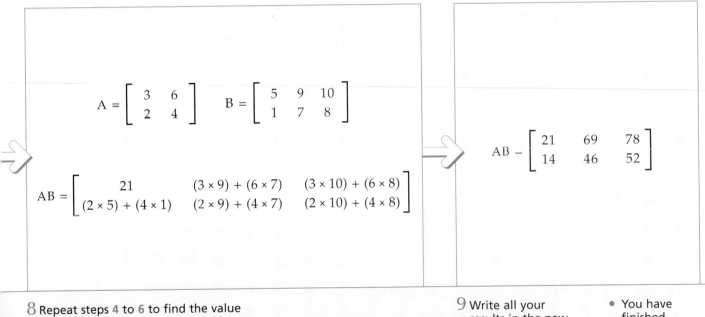

$$A = \begin{bmatrix} 3 & 6 \\ 2 & 4 \end{bmatrix} \quad B = \begin{bmatrix} 5 & 9 & 10 \\ 1 & 7 & 8 \end{bmatrix}$$

$$AB = \begin{bmatrix} 21 & (3 \times 9) + (6 \times 7) & (3 \times 10) + (6 \times 8) \\ (2 \times 5) + (4 \times 1) & (2 \times 9) + (4 \times 7) & (2 \times 10) + (4 \times 8) \end{bmatrix}$$

$$AB = \begin{bmatrix} 21 & 69 & 78 \\ 14 & 46 & 52 \end{bmatrix}$$

8 Repeat steps 4 to 6 to find the value of each number in the new matrix.

9 Write all your results in the new matrix, replacing the x variables and the subscripts.

• You have finished multiplying the matrices.

Working With Matrices

Question 1. Determine the order of each of the following matrices.

$$A = \begin{bmatrix} 1 & -4 & 8 \\ 2 & 10 & -6 \end{bmatrix} \qquad B = \begin{bmatrix} 4 & -20 \\ -1 & 0 \\ 0 & 0 \end{bmatrix} \qquad C = \begin{bmatrix} a & b \\ c & d \end{bmatrix}$$

Question 2. Add or subtract the following matrices, if possible, to determine the matrices A + B, A + C, C + D, B − A and D − B.

$$A = \begin{bmatrix} 2 & 4 \\ -3 & 5 \end{bmatrix} \qquad B = \begin{bmatrix} -3 & 10 \\ 0 & 2 \end{bmatrix}$$

$$C = \begin{bmatrix} 1 & 1 & 0 \\ 2 & -3 & 8 \end{bmatrix} \qquad D = \begin{bmatrix} 2 & -1 & 7 \\ 4 & -9 & 11 \end{bmatrix}$$

Question 3. If A is a 2 x 2 matrix, B is a 3 x 3 matrix, C is a 3 x 2 matrix and D is a 2 x 3 matrix, determine whether you can multiply the following matrices and, if so, give the order of the resulting matrix.

a) AD

b) AB

c) CD

d) DC

e) DA

f) DB

Question 4. Multiply the following matrices by a scalar to determine
the matrices 2A, −3B, 10C and 2B.

$$A = \begin{bmatrix} 2 & 3 \\ -1 & 4 \\ -2 & 5 \end{bmatrix} \qquad B = \begin{bmatrix} 0 & -4 \\ 1 & 1 \\ 2 & -3 \end{bmatrix} \qquad C = \begin{bmatrix} 3 & 10 \\ -2 & 5 \end{bmatrix}$$

Question 5. Multiply the following matrices, if possible, to
determine the matrices AB, BA, AC, BD, CD and DC.

$$A = \begin{bmatrix} 0 & 2 \\ 0 & -3 \end{bmatrix} \qquad B = \begin{bmatrix} 5 & 7 \\ 0 & 0 \end{bmatrix}$$

$$C = \begin{bmatrix} 1 & -2 & 3 \\ 0 & 2 & 5 \end{bmatrix} \qquad D = \begin{bmatrix} 0 & 1 & 4 \\ -6 & 2 & -5 \\ 1 & 1 & 0 \end{bmatrix}$$

*You can check your
answers on page 274.*

137

Chapter 7

What could be more fun than solving an equation? Solving two equations, of course. Finding the mysterious values that will solve two equations might seem like a daunting task, but the methods covered in Chapter 7 make it a snap.

Solving Systems of Equations

In this Chapter...

Solve a System of Equations by Graphing

A system of equations is a set of two or more equations. The lines representing the equations only cross each other once, unless the lines are parallel. When you find the location in the coordinate plane where the lines intersect, you find the ordered pair, or values for x and y, that makes each of the equations true. Finding the ordered pair is referred to as solving the system of equations.

A simple method for solving a system of equations is to graph the equations in a coordinate plane and note the point, or ordered pair, where the lines cross. Solving a system of equations by graphing, however, can make it difficult to obtain the correct answer. You must make sure your graphs are extremely accurate. Also, if the ordered pair contains a fraction, the fraction may be difficult to determine from the graph. The nature of the graphing method makes it critical to check your answers. To verify your solution, plug the coordinates into both equations to make sure that each equation works out correctly.

When you solve a system of equations by graphing, if you find that the two lines do not intersect, the lines are most likely parallel and there is no solution to the system of equations.

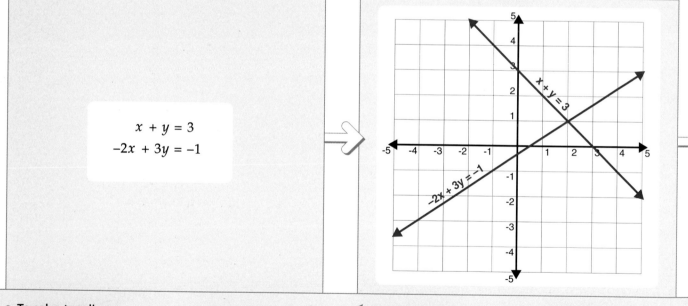

$$x + y = 3$$
$$-2x + 3y = -1$$

- To solve two linear equations, known as a system of equations, you need to find the point where the two lines intersect. The point of intersection identifies the value of each variable that solves both equations.

Note: A system of equations is a group of two or more equations.

1 To solve a system of equations by graphing the equations, start by graphing the first equation in a coordinate plane.

Note: To graph an equation in a coordinate plane, see page 84.

2 Graph the second equation in the same coordinate plane.

Tip

How can I use systems of equations?

You can use systems of equations to help organize information and solve many word problems you encounter. For instance, if Emily is four times as old as Brooke and between them, they have lived 20 years combined, how old are Emily and Brooke? You can create and solve a system of equations to determine the age of each girl. Let x represent Emily's age and y represent Brooke's age.

$$x = 4y$$
$$x + y = 20$$

Practice

$a + b = c$

Solve the following systems of equations by graphing. You can check your answers on page 259.

1) $x + 2y = 3$, $x - y = 0$
2) $2x - 3y = -6$, $-x + 4y = 8$
3) $-x - y = -1$, $4x + 3y = 2$
4) $3x - y = -2$, $x + 5y = -6$
5) $4x + 5y = 0$, $x - y = 0$
6) $3x + y = 3$, $x - 2y = 8$

$x + y = 3$

$-2x + 3y = -1$

(2,1)

Let $x = 2$, $y = 1$

$$x + y = 3$$
$$2 + 1 = 3$$
$$3 = 3 \quad \text{Correct!}$$

Let $x = 2$, $y = 1$

$$-2x + 3y = -1$$
$$-2(2) + 3(1) = -1$$
$$-4 + 3 = -1$$
$$-1 = -1 \quad \text{Correct!}$$

3 Find the point where the two lines intersect.

4 Write the point as an ordered pair in the form (x,y). The ordered pair indicates the x and y value that solves both equations.

Note: An ordered pair is two numbers, written as (x,y), that give the location of a point in the coordinate plane.

- In this example, the lines intersect at the point $(2,1)$.

5 To check your answer place the numbers you found into both of the original equations and solve the problems. In each equation, if both sides of the equation are equal, you have correctly solved the problem.

- In this example, the solution to the system of equations is $(2,1)$ or $x = 2$, $y = 1$.

Solve a System of Equations by Substitution

A system of equations is a group of two or more equations. The equations in a system of equations usually have an x and a y value in common. Finding the x and y values, called an ordered pair, that make each of the equations true is referred to as solving the system of equations.

One of the most useful techniques that can be used to solve a system of equations is called the substitution method. The substitution method allows you to solve for a variable in one of the equations and then use that solution to determine the values of the variables in the other equation. The resulting ordered pair is the solution for the system of equations.

As you substitute back and forth between the two equations, this method may seem like a sleight of hand, but the substitution method will solve every system of equations that has a solution, no matter how complicated. Due to the potential for errors in all of the substitutions, however, it is always wise to check your work.

$$x + 4y = 11$$
$$-4x + 7y = 2$$

Solve for x.

$$x + 4y = 11$$
$$x + 4y - 4y = 11 - 4y$$
$$x = 11 - 4y$$

Find the value of y.
Let $x = 11 - 4y$.

$$-4x + 7y = 2$$
$$-4(11 - 4y) + 7y = 2$$
$$-44 + 16y + 7y = 2$$
$$-44 + 23y = 2$$
$$-44 + 23y + 44 = 2 + 44$$
$$\frac{23y}{23} = \frac{46}{23}$$
$$y = 2$$

- To solve two linear equations, known as a system of equations, you can use the substitution method to identify the value of each variable that solves both equations.

1 In one equation, solve for one variable, such as x.

Note: In this example, we chose to solve for x in the first equation since this appears to be the easiest variable to solve for. For information on working with equations with more than one variable, see page 70.

- In this example, subtract $4y$ from both sides of the first equation to solve for x.

2 In the other equation, replace the variable you just solved for with the result you determined in step 1.

- In this example, replace x in the equation $-4x + 7y = 2$ with $11 - 4y$.

3 Solve the equation for the other variable, such as y.

- In this example, simplify each side of the equation. Then add 44 to both sides of the equation and then divide both sides of the equation by 23.

Note: For information on solving equations, see page 68.

Tip

What common mistake can I avoid when using the substitution method?

After you have solved for a variable in one of the equations, make sure you use the solution you found in the other equation, not the equation you just solved. If you mistakenly use your solution in the same equation that you just solved, all the variables will cancel each other out and you will not be able to determine the value of the other variable.

Practice

$a+b=c$

Solve the following systems of equations by using the substitution method. You can check your answers on page 260.

1) $2x + y = 2$, $3x - y = 3$
2) $3y - x = 7$, $x + y = 1$
3) $2x - 3y = -2$, $y + 3x = 8$
4) $2x + 4y = -x - y$, $x + y + 2 = 2$
5) $2x = 4y + 8$, $y = 3x + 3$
6) $x + y = 2y - x + 2$, $x - 2y = 4$

Find the value of x.
Let $y = 2$.

$$x + 4y = 11$$
$$x + 4(2) = 11$$
$$x + 8 = 11$$
$$x + 8 - 8 = 11 - 8$$
$$x - 3$$

Let $x = 3$ and $y = 2$

$$x + 4y = 11$$
$$3 + 4(2) = 11$$
$$3 + 8 = 11$$
$$11 = 11 \quad Correct!$$

Let $x = 3$ and $y = 2$

$$-4x + 7y = 2$$
$$-4(3) + 7(2) = 2$$
$$-12 + 14 = 2$$
$$2 = 2 \quad Correct!$$

---

4 Place the value of the variable you determined in step **3** into either of the original equations.

- In this example, let y equal 2 in the equation $x + 4y = 11$.

5 Solve the problem to determine the value of the other variable.

- In this example, subtract 8 from both sides of the equation to solve the problem.

6 To check your answer, place the numbers you found into both of the original equations and solve the problems. For each equation, if both sides of the equation are equal, you have correctly solved the problem.

- In this example, let x equal 3 and let y equal 2. Since these numbers correctly solve both equations, the solution to the system of equations is (3,2) or $x = 3$, $y = 2$.

Solve a System of Equations by Elimination

A system of equations is a set of two or more equations. The equations in a system of equations usually have an x and a y value in common. Finding the x and y values, called an ordered pair, that make each of the equations true is referred to as solving the system of equations.

The elimination method is a useful technique for solving a system of equations. The idea behind this technique is to eliminate one of the variables by canceling it out. There are several different methods of accomplishing this. The method you

choose depends on how the equations relate to one another.

When you encounter equations that share variables with the same coefficient, or number in front of the variables, but a different sign (+ or –), you can simply add one equation to the other equation to eliminate the variable and create a solution for the one variable that is left. For example, if you have $2x - 5y = 10$ and $-2x + 7y = 5$, you would add the equations. The $2x$ and $-2x$ cancel each other out, allowing you to solve for y in the resulting equation.

Simple Eliminations

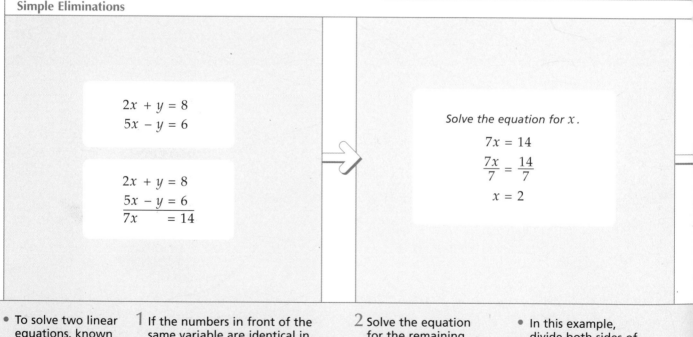

- To solve two linear equations, known as a system of equations, you can use the elimination method to identify the value of each variable that solves both equations.

1 If the numbers in front of the same variable are identical in two equations, but have a different sign (+ or –), add the equations together to eliminate the variable and solve the equations.

Note: When a number does not appear in front of a variable, assume the number is 1. For example, y equals $1y$ and $-y$ equals $-1y$.

2 Solve the equation for the remaining variable, such as x.

- In this example, divide both sides of the equation by 7 to place the x variable by itself on one side of the equation.

Note: For information on solving equations, see pages 64 to 67.

Important!

What should I do if the numbers in front of the same variable are identical, but they also have the same sign (+ or –)?

Instead of adding the equations, you can subtract one equation from the other equation to eliminate the variable. Follow the steps described below, except subtract the equations in step 1 rather than adding the equations.

$$6x + 3y = 15$$
$$- \quad 4x + 3y = 7$$
$$\overline{\quad 2x \qquad = 8}$$

Practice

$a + b = c$

Solve the following systems of equations by using the elimination method. You can check your answers on page 260.

1) $x + y = 2$, $x - y = 4$

2) $2x + 5y = 6$, $-2x - 4y = 2$

3) $x + 3y = 0$, $2x - 3y = 9$

4) $x + y = 3$, $x + 2y = 5$

5) $4x - y = 2$, $2x - y = 4$

6) $2x + 3y = -1$, $5x + 3y = 5$

Solve the equation for y.

Let $x = 2$

$2x + y = 8$

$2(2) + y = 8$

$4 + y = 8$

$4 + y - 4 = 8 - 4$

$y = 4$

Let $x = 2$ and $y = 4$

$2x + y = 8$

$2(2) + 4 = 8$

$4 + 4 = 8$

$8 = 8$ Correct!

Let $x = 2$ and $y = 4$

$5x - y = 6$

$5(2) - 4 = 6$

$10 - 4 = 6$

$6 = 6$ Correct!

---

3 To solve the equation for the other variable, place the value of the variable you determined in step **2** into either one of the original equations.

- In this example, let x equal 2 in the equation $2x + y = 8$.

4 Solve the equation to determine the value of the variable.

- In this example, subtract 4 from both sides of the equation to place the y variable by itself on one side of the equation.

5 To check your answer, place the numbers you found into both of the original equations and solve the problems. For each equation, if both sides of the equation are equal, you have correctly solved the problem.

- In this example, let x equal 2 and let y equal 4. Since these numbers correctly solve both equations, the solution to the system of equations is (2,4) or $x = 2$, $y = 4$.

CONTINUED

145

Solve a System of Equations by Elimination *continued*

Solving a system of equations involves finding the x and y values, or ordered pair, that will make each equation in a group of equations true. The elimination method allows you to cancel out one of the variables and solve a system of equations.

Often, the variables in each equation have different coefficients, or numbers in front of the variables. In this instance, you have to multiply both equations so that either the x or y variable in both equations ends up with the same coefficient. Before you multiply the equations, you must first choose which variable you want to eliminate,

keeping in mind that smaller numbers in front of variables are easier to work with. Then multiply the first equation by the coefficient of the variable you want to eliminate in the second equation and multiply the second equation by the coefficient of the variable you want to eliminate in the first equation. The resulting equations will then have a set of matching variables that you can eliminate by adding or subtracting the equations. Once the variable is eliminated, you will be able to solve for the other variable and work out a solution for the system of equations.

More Challenging Eliminations

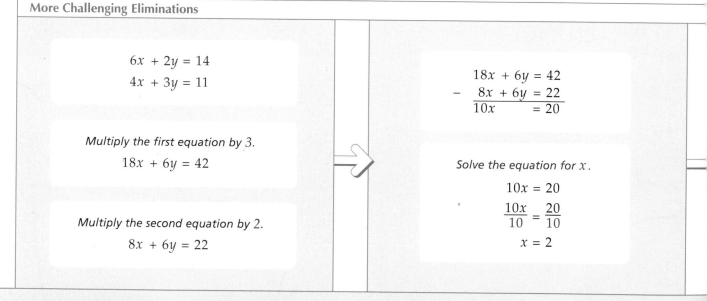

$$6x + 2y = 14$$
$$4x + 3y = 11$$

Multiply the first equation by 3.
$$18x + 6y = 42$$

Multiply the second equation by 2.
$$8x + 6y = 22$$

$$\begin{array}{r} 18x + 6y = 42 \\ -\ \underline{8x + 6y = 22} \\ 10x\qquad\ = 20 \end{array}$$

Solve the equation for x.
$$10x = 20$$
$$\frac{10x}{10} = \frac{20}{10}$$
$$x = 2$$

- If the numbers in front of the same variable are not identical in two equations, you will first need to change both equations to eliminate the variable and solve the equations.

- In this example, we want to eliminate the y variable.

1 Multiply each term in the first equation by the number in front of the variable you want to eliminate in the second equation.

2 Multiply each term in the second equation by the number in front of the variable you want to eliminate in the first equation.

3 Add or subtract the two equations. In this example, we subtract the equations.

Note: If the same sign (+ or −) appears before the numbers in front of the variables you want to eliminate, subtract the equations. If different signs appear before the numbers in front of the variables you want to eliminate, add the equations.

4 Solve the equation for the remaining variable, such as x.

- In this example, divide both sides of the equation by 10.

Note: For information on solving equations, see pages 64 to 67.

Tip

Is there a faster way to eliminate a variable instead of multiplying both equations?

If the number in front of a variable evenly divides into the number in front of the same variable in the other equation, you can simply multiply each term in the first equation by the number of times the smaller number divides into the larger number. For example, in the equations $2x + 3y = 5$ and $4x + 5y = 11$, 2 divides into 4 twice, so you could multiply each term in the first equation by 2 to make the number in front of the x variable in both equations equal to 4. In this example, $2x + 3y = 5$ would end up as $4x + 6y = 10$. You could then subtract the equations to eliminate the x variable.

Practice

$a + b = c$

Solve the following systems of equations by using the elimination method. You can check your answers on page 260.

1) $2x + y = 1$, $3x + 2y = 1$
2) $x - y = 0$, $2x + 3y = 0$
3) $y - 2x = -2$, $3x - 4y = -2$
4) $-3x + 2y = 7$, $3x - 5y = -13$
5) $x + y = 4$, $2x - 4y = 2$
6) $7x + y = 9$, $4x - 3y = -2$

Solve the equation for y.

Let $x = 2$

$$6x + 2y = 14$$
$$6(2) + 2y = 14$$
$$12 + 2y = 14$$
$$12 + 2y - 12 = 14 - 12$$
$$\frac{2y}{2} = \frac{2}{2}$$
$$y = 1$$

Let $x = 2$ and $y = 1$

$$6x + 2y = 14$$
$$6(2) + 2(1) = 14$$
$$12 + 2 = 14$$
$$14 = 14 \quad Correct!$$

Let $x = 2$ and $y = 1$

$$4x + 3y = 11$$
$$4(2) + 3(1) = 11$$
$$8 + 3 = 11$$
$$11 = 11 \quad Correct!$$

---

5 To solve the equation for the other variable, place the value of the variable you determined in step 4 into either one of the original equations.

- In this example, let x equal 2 in the equation $6x + 2y = 14$.

6 Solve the equation to determine the value of the variable.

- In this example, subtract 12 from both sides of the equation and then divide both sides of the equation by 2.

7 To check your answer, place the numbers you found into both of the original equations and solve the problems. For each equation, if both sides of the equation are equal, you have correctly solved the problem.

- In this example, let x equal 2 and let y equal 1. Since these numbers correctly solve both equations, the solution to the system of equations is $(2,1)$ or $x = 2$, $y = 1$.

Solving Systems of Equations

Question 1. Solve the following systems of equations by graphing.

    a) $2x + 3y = 0$
       $3x - y = 0$

    b) $x - y = 0$
       $2x + 3y = 5$

    c) $x + 4y = 7$
       $2x - 3y = -8$

    d) $3x + 5y = -3$
       $-2x + 6y = 2$

    e) $-x - y = 0$
       $3x + y = 2$

Question 2. Solve the following systems of equations by substitution.

    a) $x + y = 7$
       $x - y = -3$

    b) $-10x - 5y = 0$
       $-2x + y = 0$

    c) $2x - 3y = 9$
       $-x + 2y = -5$

    d) $2x + 3y = -10$
       $x - 5y = 8$

    e) $2x - y = 2$
       $10x + 5y = 10$

Question 3. Solve the following systems of equations by elimination.

a) $x - y = 0$
$2x + 3y = 5$

b) $3x + 2y = 12$
$x - y = -1$

c) $7x - 4y = 4$
$11x + 2y = -2$

d) $x + 3y = -1$
$-x + 2y = -4$

e) $6x - 2y = 12$
$-5x + y = -12$

Question 4. Solve the following systems of equations by any method.

a) $2x + 3y + 5 = 4$
$x - 4y + 7 = 12$

b) $10x - 10y + 2 = 3y - 20$
$5 + y = x + 6$

c) $4y + 3x = 7y$
$5x - y = -2x - y - 7$

d) $5x - 2y + 8 = 6y + 6x + 7$
$12x - 10 = 4y + 2$

e) $3y - x + 1 = -5 - y$
$x + y = y - 2$

*You can check your
answers on page 275.*

149

Chapter 8

While the name makes polynomials sound like an obscure shape from geometry class, polynomials are actually simple mathematical statements with an interesting name. Chapter 8 introduces you to the different types of polynomials and gives you a few tried-and-true techniques for working with them.

Working with Polynomials

In this Chapter...

Introduction to Polynomials

A polynomial is an expression that contains one or more numbers and variables and involves only addition, subtraction and multiplication. Polynomials do not include complicated operations like division, square roots or absolute values. A polynomial can be simple, like $7a$, or complicated, like $2a^3 + 3b^2 - c + 4d$.

Polynomials are often classified by the number of terms they contain. For example, a polynomial with one term, such as $3x$, is called a monomial.

Polynomials can also be classified by degree, which indicates the highest exponent to which the polynomial's terms are raised. For example, a polynomial with a degree of 1 is called a linear polynomial. The terms in polynomials are generally written in order from highest exponential value to lowest exponential value.

Classifications can be combined to better describe a polynomial. For example, $3x^2 + 5y$ could be classified as a quadratic binomial, since it has a degree of 2 and contains two terms.

About Polynomials

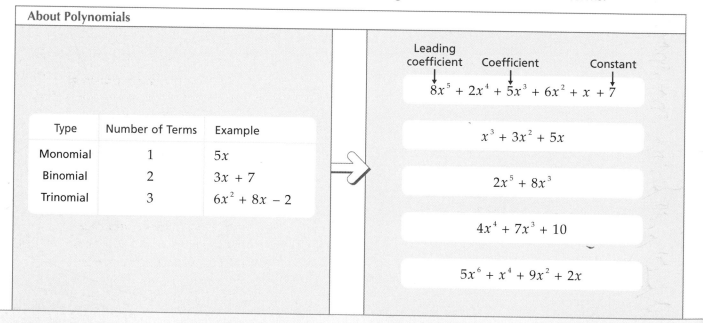

Type	Number of Terms	Example
Monomial	1	$5x$
Binomial	2	$3x + 7$
Trinomial	3	$6x^2 + 8x - 2$

Leading coefficient, Coefficient, Constant

$$8x^5 + 2x^4 + 5x^3 + 6x^2 + x + 7$$

$$x^3 + 3x^2 + 5x$$

$$2x^5 + 8x^3$$

$$4x^4 + 7x^3 + 10$$

$$5x^6 + x^4 + 9x^2 + 2x$$

- A polynomial consists of one or more terms, which can be a combination of numbers and/or variables, that are added together or subtracted from one another.

- A polynomial with one term is called a monomial. A polynomial with two terms is called a binomial. A polynomial with three terms is called a trinomial.

  Note: Polynomials with four or more terms do not have a special name—just call them polynomials.

- Polynomials are usually written with the terms listed in descending order, from the term with the highest exponent to the term with the lowest exponent.

- The number in front of a variable is called the coefficient.

- The number in front of the variable with the highest exponent is called the leading coefficient.

- A term that contains only a number is called a constant.

Tip

If a polynomial contains more than one variable, how can I determine the degree of the polynomial?

If you are asked to determine the degree of a polynomial such as $4x^3y^2 + 3xy$, you can add the exponents in each term together. The term with the highest total of exponents will be the degree of the polynomial. For instance, in the polynomial $4x^3y^2 + 3xy$, the degree of the polynomial is 5 (quintic) since $3 + 2$ equals 5. Remember, also, that when a variable does not show an exponent, you can assume the exponent is 1.

Practice

Describe each of the following polynomials based on the degree of the polynomial and the number of terms in the polynomial. You can check your answers on page 260.

1) 14

2) $x^3 - 6$

3) $x + y - 2z - 8$

4) $x^2 + y^2 - z + 2x + 4$

5) $abc^2 + a^3 - 2bc$

6) $3a^5 + 2a^3b^2$

The Degree of a Polynomial

Degree	Name	Example
0	constant	6
1	linear	$4x + 8$
2	quadratic	$3x^2 + 11x + 7$
3	cubic	$x^3 + 15x + 2$
4	quartic	$5x^4 + 6x^2 + 9$
5	quintic	$x^5 + x^3 + x^2 + x$

Describing a Polynomial

3	constant monomial
$4x$	linear monomial
$9x^2 - 2x$	quadratic binomial
$5x^3 + 11$	cubic binomial
$7x^4 + 6x^3 - 4x$	quartic trinomial
$8x^5 - 3x^3 + 9$	quintic trinomial

- The highest exponent in a polynomial is called the degree of the polynomial.

- Each degree has a name. For example, a polynomial with a degree of 2 is called a quadratic polynomial.

- The degrees shown in the above list are the most commonly found in algebra.

- You can describe a polynomial based on the number of terms in the polynomial and the degree, or highest exponent in the polynomial.

- When describing a polynomial, write the name of the degree followed by the word used to describe the number of terms.

- For example, if the degree of a polynomial is three and the polynomial has two terms, write "cubic binomial."

Multiply Polynomials

A polynomial is an expression with one or more variables combined together, like $3x^3 + 5y^4$. A polynomial with only one term, like $7a^2$, is called a monomial. Multiplying, or distributing, polynomials can seem intimidating but is actually a simple concept that will become second nature after a bit of practice.

There are no rules restricting which kinds of polynomials can be multiplied together—the two polynomials do not have to contain the same variables or even the same number of terms.

To multiply two polynomials you methodically multiply each term of the first polynomial by each term of the second polynomial. You simply apply the distributive property using each term of the first polynomial (see page 30). Multiplying a monomial by a polynomial follows the same method—just multiply the monomial's single term by each of the terms contained in the polynomial. After all of the multiplying, you should simplify the resulting numbers and variables by adding them together.

Multiplying Polynomials Example 1

Polynomial   Polynomial

$$(x + 2)(y - 3)$$

$$(x)(y) + (x)(-3) + (2)(y) + (2)(-3)$$

$$xy + (-3x) + 2y + (-6)$$

$$xy - 3x + 2y - 6$$

Multiplying Polynomials Example 2

$$(x + 4y)(x - 2y)$$

$$(x)(x) + (x)(-2y) + (4y)(x) + (4y)(-2y)$$

$$x^2 + (-2xy) + 4xy + (-8y^2)$$

$$x^2 + 2xy + (-8y^2)$$

$$x^2 + 2xy - 8y^2$$

- You can multiply polynomials together. A polynomial consists of one or more terms, which can include numbers and/or variables.

1 To multiply two polynomials together, multiply each term in the first polynomial by each term in the second polynomial. Then add the results together.

Note: When multiplying terms, placing the terms in parentheses ( ) can help prevent errors with the signs (+ or −) in front of the terms.

2 To simplify your result, add and subtract any numbers and/or variables that can be combined.

1 To multiply two polynomials together, multiply each term in the first polynomial by each term in the second polynomial. Then add the results together.

Note: When you multiply a variable by itself, you can use exponents to simplify the problem. For example, $x \times x$ equals x^2. For more information on exponents, see page 54.

2 To simplify the result, add and subtract any numbers and/or variables that can be combined.

Tip

After multiplying polynomials, how can I better present my answer?

Listing the terms in your answer in some sort of order will help simplify your answer. If you have only one variable in your answer, list the terms in order of descending exponent, such as $x^5 + x^3 + x^2 + x$. If you have more than one variable in your answer, list the variables from highest to lowest total of exponents of the variables. For example, the exponent total of x^3y^2 is 5 while the exponent total of x^2y^2 is 4, so you would list the variables as $x^3y^2 + x^2y^2 + xy^2 + x$.

Practice

a + b = c

Multiply the following polynomials and simplify your answers. You can check your answers on page 260.

1) $(x + 4)(x + 2)$

2) $3x^4(x^2 - x + 2)$

3) $(x + 1)(y - 1)$

4) $x^2(x + y - 5)$

5) $(x + y)(2x + 3y)$

6) $(x + y)(x - y)$

Multiplying Polynomials Example 3

$$(x + 3y)(2x + 4x - 5y)$$

$$(x)(2x) + (x)(4x) + (x)(-5y) + (3y)(2x)$$
$$+ (3y)(4x) + (3y)(-5y)$$

$$2x^2 + 4x^2 + (-5xy) + 6xy + 12xy + (-15y^2)$$

$$6x^2 + 13xy - 15y^2$$

1 To multiply two polynomials together, multiply each term in the first polynomial by each term in the second polynomial. Then add the results together.

2 To simplify the result, add and subtract any numbers and/or variables that can be combined.

Multiplying Polynomials Example 4

$$(a - 2b)(3a^2 + 4ab - 5b^2)$$

$$(a)(3a^2) + (a)(4ab) + (a)(-5b^2) + (-2b)(3a^2)$$
$$+ (-2b)(4ab) + (-2b)(-5b^2)$$

$$3a^3 + 4a^2b + (-5ab^2) + (-6a^2b) + (-8ab^2) + 10b^3$$

$$3a^3 + (-2a^2b) + (-13ab^2) + 10b^3$$

$$3a^3 - 2a^2b - 13ab^2 + 10b^3$$

1 To multiply two polynomials together, multiply each term in the first polynomial by each term in the second polynomial. Then add the results together.

Note: When multiplying variables with exponents that have the same letter, you can add the exponents. For example, $a^2 \times a^2$ equals a^{2+2}, which equals a^4. If an exponent does not appear beside a variable, you can assume the exponent is 1. For example, a is the same as a^1.

2 To simplify the result, add and subtract any numbers and/or variables that can be combined.

Divide Polynomials Using Long Division

Think back to the days before you used a calculator to figure out division problems. Remember those long calculations on the blackboard? While that may be just a foggy memory, you are going to pull those long division skills out of retirement to divide polynomials.

First, you should brush up on some notation and terminology. If you are given the problem $x \div y$, the problem is written as $y\overline{)x}$ when you want to work it out using long division. The expression being divided, represented by x in this case, is called the dividend and the expression doing the dividing, represented by y in this case, is known as the divisor. The answer is called the quotient.

For things to work out neatly, the divisor and dividend expressions should both have the variable's exponents appearing in descending order, such as x^3, x^2, x. If an exponent is missing, you should add a term with that exponent to the problem, with a 0 before the term, such as $0x^2$. The term will simply act as a placeholder to help line everything up.

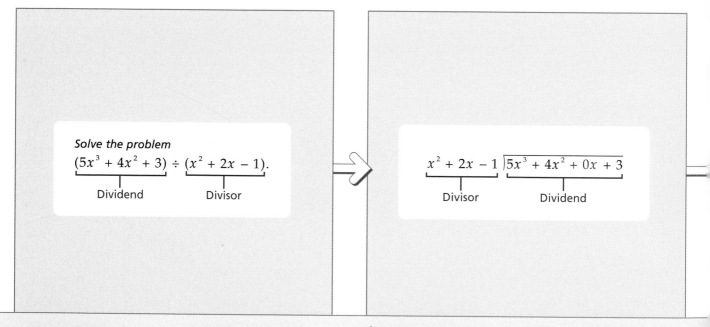

Solve the problem
$$(5x^3 + 4x^2 + 3) \div (x^2 + 2x - 1).$$
Dividend   Divisor

$$x^2 + 2x - 1 \overline{)5x^3 + 4x^2 + 0x + 3}$$
Divisor   Dividend

- You can divide one polynomial by another polynomial using long division.

  Note: A polynomial consists of one or more terms, which can be a combination of numbers and/or variables, that are added together or subtracted from one another.

- In a division problem, the polynomial you are dividing into is known as the dividend. The polynomial you are dividing by is known as the divisor.

1 To divide polynomials, rewrite the division problem in long division notation. Make sure you place the divisor to the left of the division symbol and the dividend below the division symbol.

2 In the divisor and dividend, make sure the variable's exponents appear in descending order, such as x^3, x^2, x. If an exponent is missing, add a term with that exponent to the problem, with a 0 before the term.

- In this example, the x term was missing in the dividend, so the $0x$ term was added.

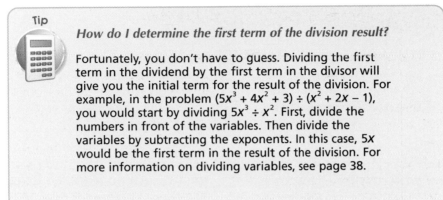

Tip

How do I determine the first term of the division result?

Fortunately, you don't have to guess. Dividing the first term in the dividend by the first term in the divisor will give you the initial term for the result of the division. For example, in the problem $(5x^3 + 4x^2 + 3) \div (x^2 + 2x - 1)$, you would start by dividing $5x^3 \div x^2$. First, divide the numbers in front of the variables. Then divide the variables by subtracting the exponents. In this case, $5x$ would be the first term in the result of the division. For more information on dividing variables, see page 38.

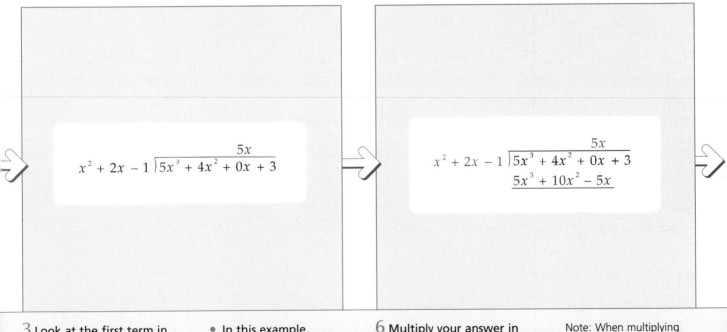

3 Look at the first term in the divisor and the first term in the dividend.

4 Determine what you need to multiply by the first term in the divisor to give you the first term in the dividend.

- In this example, $x^2 \times 5x$ equals $5x^3$.

5 Write your answer above the division symbol, directly above the similar term in the dividend.

6 Multiply your answer in step 5 by each term in the divisor and write the result under the dividend, directly below the similar terms.

- In this example, multiply $5x \times (x^2 + 2x - 1)$.

Note: When multiplying variables with exponents that have the same letter, add the exponents. For example, $x^2 \times x$ equals x^{2+1}, which equals x^3.

7 Draw a horizontal line directly below your answer in step 6.

CONTINUED

Divide Polynomials Using Long Division *continued*

While there are a lot of steps in the process of using long division to divide polynomials, try not to be intimidated. If you work methodically, the process is not difficult and should always provide you with a solution.

As you work through the process of long division, you will know that you have found the quotient when you have no more terms to work with in the dividend. If the dividend polynomial has divided evenly, you will have no terms left at the bottom of your stack of calculations. If there

does happen to be a term or two hanging around the bottom of your work, you simply count these terms as your remainder. You list these remainder terms next to the quotient, as a fraction of the divisor.

Admittedly, using long division to divide polynomials can be initially confusing. As is the case with so many mathematical methods, however, becoming proficient is just a matter of practicing until you are completely comfortable with the process.

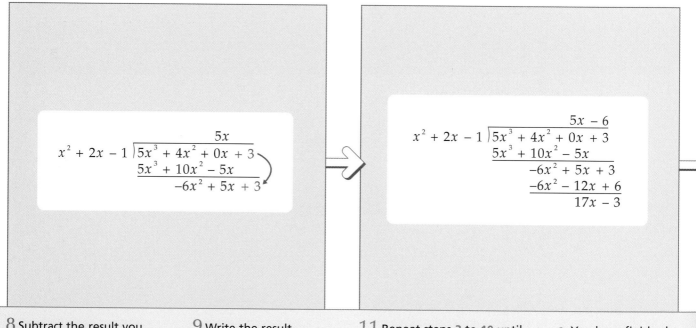

8 Subtract the result you determined in step **6** from the terms that appear directly above. The first term should cancel itself out.

Note: When subtracting terms, if you have two negative (−) signs beside one another, you can replace the signs with one plus (+) sign. For example, $0x - (-5x)$ equals $0x + 5x$.

9 Write the result below the horizontal line.

10 Write the next term in the dividend at the end of the result. In this example, write + 3 at the end of the result.

11 Repeat steps **3** to **10** until there are no more terms in the dividend in step **10**. In step **3**, make sure you look at the first term in the divisor and the first term in the last line below the dividend.

• You have finished dividing the polynomials.

Practice
$a+b=c$

Divide the following polynomials using long division.
You can check your answers on page 260.

1) $x^2 + x + 1 \overline{\smash{)}-x^3 + 2x^2 + 3x + 4}$

2) $x^2 - 4 \overline{\smash{)}3x^4 - 12x^2 - 4x}$

3) $3x + 2 \overline{\smash{)}6x^4 + 4x^3 - 2}$

4) $x^2 + 1 \overline{\smash{)}x^5 + 2x^3 + x}$

5) $x^3 + 2x + 1 \overline{\smash{)}x^7 + 2x^6 + 2x^4 + x^3 + 2x^2 + 3x + 1}$

6) $x + 1 \overline{\smash{)}x^2}$

$$5x - 6 + \frac{17x - 3}{x^2 + 2x - 1}$$

Check Your Answer

$(5x - 6)(x^2 + 2x - 1) + 17x - 3$
$5x^3 + 10x^2 - 5x - 6x^2 - 12x + 6 + 17x - 3$
$5x^3 + 4x^2 + 3$ *Correct!*

12 To write your answer, write the polynomial above the division symbol. Then add a fraction, with the remainder, or left over, as the top part of the fraction and the original divisor as the bottom part of the fraction.

1 To check your answer, multiply the polynomial above the division symbol by the original divisor. Then add the result to the remainder.

- In this example, the polynomial above the division symbol is $(5x - 6)$, the original divisor is $(x^2 + 2x - 1)$, and the remainder is $17x - 3$.

Note: For more information on multiplying polynomials, see page 154.

- If you end up with the original dividend, you have correctly solved the problem.

Divide Polynomials Using Synthetic Division

If dividing polynomials using long division (see page 156) tests the limits of your endurance, you are going to love synthetic division. Synthetic division is an easy method of dividing polynomials, but this method can only be used when you are faced with the task of dividing a polynomial by a linear polynomial in the form of $x - c$, where c is a number.

Synthetic division is also a faster method of dividing polynomials than the long division method because you perform all of the required calculations with the coefficients, or the numbers in front of the variables, and leave all of the variables off to the side.

On paper, synthetic division looks strange and complicated. In reality, though, the process is very straightforward, amounting to nothing more than a few quick addition and multiplication calculations.

In a division problem, the polynomial you are dividing into is known as the dividend. The polynomial you are dividing by is known as the divisor. Before you begin dividing, make sure the variable's exponents in the dividend appear in descending order, such as x^3, x^2, x. If an exponent is missing, add a term with that exponent to the problem, with a 0 before the term, such as 0x.

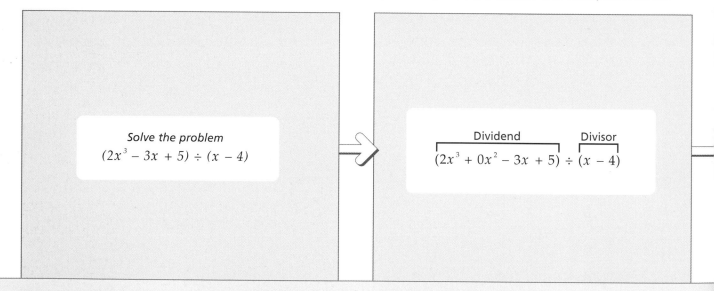

Solve the problem
$$(2x^3 - 3x + 5) \div (x - 4)$$

Dividend   Divisor
$$(2x^3 + 0x^2 - 3x + 5) \div (x - 4)$$

- Synthetic division is a shortcut method you can use when dividing polynomials.

  Note: A polynomial consists of one or more terms, which can be a combination of numbers and/or variables, that are added together or subtracted from one another.

- You can only use synthetic division when you are dividing by a polynomial with only two terms in the form $x - c$, where c is a number. For example, you can divide by the polynomial $x - 4$.

- In a division problem, the polynomial you are dividing into is known as the dividend. The polynomial you are dividing by is known as the divisor.

1 In the dividend, make sure the variable's exponents appear in descending order, such as x^3, x^2, x. If an exponent is missing, add a term with that exponent to the problem, with a 0 before the term.

- In this example, the x^2 term was missing in the dividend, so the $0x^2$ term was added.

Tip

Are there other uses for synthetic division?

You can use synthetic division to help solve equations that contain variables with exponents higher than 2, such as x^3 or x^4. For example, you can use synthetic division to help you find the solutions to the equation $2x^3 + 10x^2 - 2x - 10 = 0$. For information on solving equations with exponents that are higher than 2, see page 218.

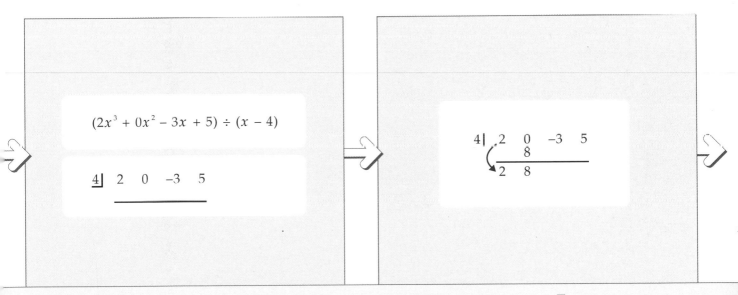

$(2x^3 + 0x^2 - 3x + 5) \div (x - 4)$

$4|\ \ 2\ \ \ 0\ \ \ -3\ \ \ 5$

$4|\ \ 2\ \ \ 0\ \ \ -3\ \ \ 5$
    8
   $2\ \ \ 8$

2 In the dividend, look at the numbers directly in front of the variables as well as the last number and write the numbers in a row.

Note: If a number does not appear in front of a variable, assume the number is 1. For example, x equals $1x$.

3 To the left of the row of numbers, write the last number in the divisor, but change the sign (+ or −) in front of the number. Draw a partial box around the number to separate the number from the other numbers.

4 Leave a blank line below the row of numbers and draw a horizontal line.

5 Drop down the first number in the dividend and write the number below the line. In this example, drop down the number 2.

6 Multiply the number in the partial box by the number below the line.

• In this example, multiply 4 by 2, which equals 8.

7 Write the result in the next column of numbers, directly above the line.

8 Add the two numbers in the second column and write the result directly under the numbers, below the line.

CONTINUED

Divide Polynomials Using Synthetic Division *continued*

As mentioned, one of the most appealing traits of synthetic division is that it allows you to work without having to deal with variables and their exponents.

When performing synthetic division, the numbers below the line indicate the result of the division, with the last number being the remainder, or left over value. If the last number is 0, the divisor evenly divides into the dividend and there is no remainder. If a remainder exists, write the remainder as a fraction, with the original divisor as the bottom part of the fraction.

Once you have your result, you can slide the variables neatly back into place, giving you the final answer.

To check your work, all you have to do is multiply the answer by the binomial that was your original divisor and then add the result to the remainder. If it works out to the polynomial you began with as your dividend, you have divided correctly.

The best way to get a firm grasp on synthetic division is to simply practice the method. After just a few tries, you will have it down pat and wonder how you ever lived without this handy technique.

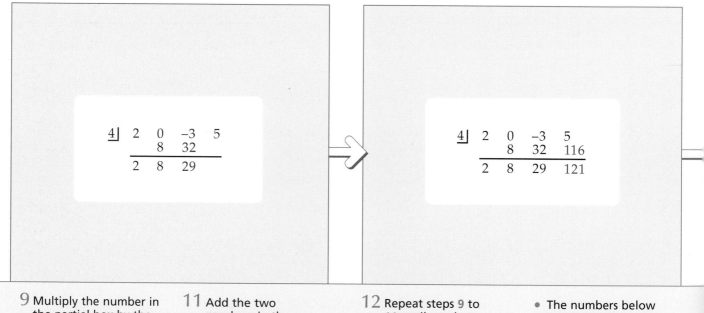

9 Multiply the number in the partial box by the last number below the line.

- In this example, multiply 4 by 8, which equals 32.

10 Write the result in the next column of numbers, directly above the line.

11 Add the two numbers in the column and write the result directly under the numbers, below the line.

- In this example, add −3 + 32, which equals 29.

12 Repeat steps **9** to **11** until you have added the two numbers in every column.

- You have finished dividing the polynomials.

- The numbers below the line indicate the result of the division, with the last number being the remainder, or left over value.

  Note: If the last number is 0, the divisor evenly divides into the dividend.

Practice

$a+b=c$

Divide the following polynomials using synthetic division. You can check your answers on page 261.

1) $(x^3 + 3x^2 - 13x + 7) \div (x - 2)$

2) $(x^4 + x^3 + 2x^2 + x) \div (x + 1)$

3) $(2x^4 + 6x^3 - x^2 - 3x + 6) \div (x + 3)$

4) $(-x^3 + x^2 + 11x + 10) \div (x + 2)$

5) $(x^6 - x^5 + x^2 + 2) \div (x - 1)$

6) $(x^3 + 5x^2 + 2x) \div (x + 5)$

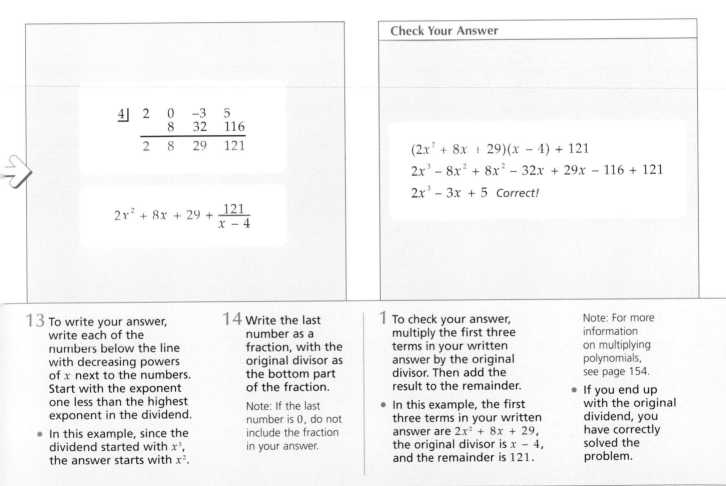

Check Your Answer

$$\underline{4|} \quad \begin{array}{cccc} 2 & 0 & -3 & 5 \\ & 8 & 32 & 116 \\ \hline 2 & 8 & 29 & 121 \end{array}$$

$2x^2 + 8x + 29 + \dfrac{121}{x - 4}$

$(2x^2 + 8x + 29)(x - 4) + 121$

$2x^3 - 8x^2 + 8x^2 - 32x + 29x - 116 + 121$

$2x^3 - 3x + 5$ *Correct!*

13 To write your answer, write each of the numbers below the line with decreasing powers of x next to the numbers. Start with the exponent one less than the highest exponent in the dividend.

- In this example, since the dividend started with x^3, the answer starts with x^2.

14 Write the last number as a fraction, with the original divisor as the bottom part of the fraction.

Note: If the last number is 0, do not include the fraction in your answer.

1 To check your answer, multiply the first three terms in your written answer by the original divisor. Then add the result to the remainder.

- In this example, the first three terms in your written answer are $2x^2 + 8x + 29$, the original divisor is $x - 4$, and the remainder is 121.

Note: For more information on multiplying polynomials, see page 154.

- If you end up with the original dividend, you have correctly solved the problem.

Working With Polynomials

Question 1. Determine the degree of the following polynomials.

a) $x^3 + x^2 + 2x + 9$

b) $x - x^4 + x^2 + 5 - 3x + 2x^4$

c) $x^5 + x^6 + x - 10$

d) $3 - 2x + 4x^2 - 5x^3 + 7x^7$

e) $x + 5$

f) $34 + 10x^2 - 7xy + 2y^2$

Question 2. Multiply the following polynomials and simplify your answers as much as possible.

a) $(x + 2)(x + 3)$

b) $(2x + 1)(x - 2)$

c) $(x - 5)(x + 5)$

d) $(x + 1)(x - 4)(2x - 3)$

e) $(x^2 + 4x)(x - 2)$

f) $(x^2 + 3x + 1)(2x^2 - x - 1)$

Question 3. Divide the following polynomials by using long division.

a) $(x^3 + x^2 - 2x - 8) \div (x - 2)$

b) $(3x^4 + x^3 - 2x^2 + x) \div (x + 1)$

c) $(-2x^3 + 3x^2 + 9x - 2) \div (2x - 1)$

d) $(2x^4 - 6x^2 - 6) \div (x^2 + 1)$

e) $(x^4 + 2x^3 - 1) \div (x^2 + x - 2)$

Question 4. Divide the following polynomials by using synthetic division.

a) $(x^3 - 1) \div (x - 1)$

b) $(2x^3 + 5x^2 - x) \div (x + 2)$

c) $(x^4 + 4x^3 - 3x^2 + 15x + 20) \div (x + 5)$

d) $(3x^4 - 11x^3 + 5x^2 + 10x - 11) \div (x - 3)$

e) $(5x^5 - 20x^4 - x^2 + 14x - 40) \div (x - 4)$

You can check your answers on page 276.

165

Chapter 9

$$\sqrt{2x + y^2}$$

$$\sqrt[3]{a^2 \cdot b^2}$$

This is not a chapter about bringing down "the establishment." Radicals might seem like outlaws at first, but they are law-abiding expressions that are straightforward and easy to work with. Chapter 9 gives you a thorough understanding of how to work with these useful little expressions.

Working with Radicals

In this Chapter...

Introduction to Radicals

Radicals are the opposite of exponents. While an exponent asks you to multiply a number by itself, radicals break a number down into equal factors. Put another way, a radical asks what number multiplied by itself a certain number of times will equal the number shown under the radical sign ($\sqrt{}$).

The radical sign ($\sqrt{}$) indicates a radical operation, which is often referred to as finding the root. The number you are trying to find the root of is located under the radical sign and is known as the radicand. Sometimes, you will also notice a small number just in front of the radical sign, such as $\sqrt[3]{}$. This number

is called the index and tells you which root you must find. For example, the expression $\sqrt[4]{16}$ asks you to find the fourth root of 16, or the number that will equal 16 when multiplied by itself four times.

The most common radical operation you will perform is the square root. When finding the square root, you are finding the number that, when multiplied by itself, will equal the radicand. Square roots are so common that if a radical does not display an index number, you can assume the index is 2 and you are looking for the square root.

About Radicals

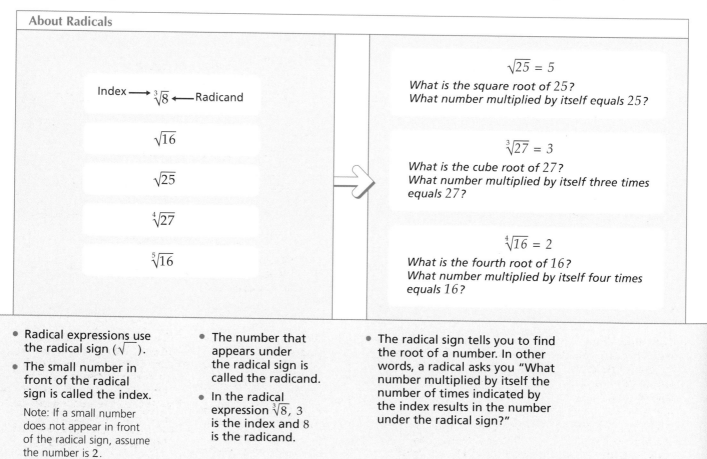

Index → $\sqrt[3]{8}$ ← Radicand

$\sqrt{16}$

$\sqrt{25}$

$\sqrt[4]{27}$

$\sqrt[5]{16}$

$\sqrt{25} = 5$

What is the square root of 25?
What number multiplied by itself equals 25?

$\sqrt[3]{27} = 3$

What is the cube root of 27?
What number multiplied by itself three times equals 27?

$\sqrt[4]{16} = 2$

What is the fourth root of 16?
What number multiplied by itself four times equals 16?

- Radical expressions use the radical sign ($\sqrt{}$).
- The small number in front of the radical sign is called the index.

  Note: If a small number does not appear in front of the radical sign, assume the number is 2.

- The number that appears under the radical sign is called the radicand.
- In the radical expression $\sqrt[3]{8}$, 3 is the index and 8 is the radicand.

- The radical sign tells you to find the root of a number. In other words, a radical asks you "What number multiplied by itself the number of times indicated by the index results in the number under the radical sign?"

Besides square roots, what other roots will I commonly find in algebra?

Aside from square roots, you will often find cube roots in algebra. Radicals with an index of 3 are called cube roots. When you are finding the cube root of a number, you are finding out what number multiplied by itself three times equals the number under the radical sign. The following are common cube roots.

$\sqrt[3]{1} = 1$ $\sqrt[3]{125} = 5$ $\sqrt[3]{512} = 8$

$\sqrt[3]{8} = 2$ $\sqrt[3]{216} = 6$ $\sqrt[3]{729} = 9$

$\sqrt[3]{27} = 3$ $\sqrt[3]{343} = 7$ $\sqrt[3]{1,000} = 10$

$\sqrt[3]{64} = 4$

Can the number under a radical sign be negative?

Yes. You can find the root of a negative number the same way you find the root of a positive number. Your answer will always be a negative number. For example, in the radical expression $\sqrt[3]{-8}$, your answer would be -2 since $(-2) \times (-2) \times (-2)$ equals -8. When the number under a radical sign is a negative number, you cannot find the root if the radical has an even-numbered index. For example, you cannot find the square root of -4, since no number can be multiplied by itself to equal a negative number.

Square Roots

$\sqrt{1} = 1$ $\sqrt{36} = 6$

$\sqrt{4} = 2$ $\sqrt{49} = 7$

$\sqrt{9} = 3$ $\sqrt{64} = 8$

$\sqrt{16} = 4$ $\sqrt{81} = 9$

$\sqrt{25} = 5$ $\sqrt{100} = 10$

Radicals with Variables

$\sqrt{x^2} = |x|$ *since* $x \times x$ *equals* x^2

What is the square root of x^2?
What variable multiplied by itself equals x^2?

$\sqrt{x^4} = x^2$ *since* $x^2 \times x^2$ *equals* x^4

What is the square root of x^4?
What variable multiplied by itself equals x^4?

$\sqrt[3]{x^3} = x$ *since* $x \times x \times x$ *equals* x^3

What is the cube root of x^3?
What variable multiplied by itself three times equals x^3?

- Radicals with an index of 2 are called square roots. You will commonly find square roots in algebra.

- The above list shows the most common square roots.

- When you need to find the square root of a number, the small number in front of the radical sign will usually not be shown, so assume the number is 2.

- When finding the square root of a number, ask yourself "What number multiplied by itself equals the number under the radical sign?"

- A variable can appear under a radical sign. As with numbers, you can find the root of variables.

  Note: When you multiply variables with the same letter, you add the exponents. For example, $x^2 \times x^2$ equals x^4.

- If the index and exponent of a variable under a radical sign match and the index is an even number, surround the variable in your answer with absolute value symbols (||). This ensures that the answer will be positive even if the variable is negative.

  Note: For more information on absolute values, see page 24.

169

Simplify Radicals

Sometimes, the best way to work with a radical is to write out the radical in its simplest form.

To simplify a radical, you remove as much as you can from beneath the radical sign. To do this, break the terms under the radical sign into their smallest parts, called factors. You will know that any numbers under the radical sign are broken down completely when all of the factors are prime numbers.

Next, if a factor appears under the radical sign the same number of times as indicated by the index, or small number in front, of the radical, you can pull those repeating factors out from under the radical sign, placing one of them to the left of the radical sign. For example, if you have $\sqrt[3]{5x^3}$, you can break x^3 down into the factors $x \times x \times x$ and then place a single x outside of the radical symbol, giving you $x\sqrt[3]{5}$.

If the exponent of the term that you remove is a multiple of the index, and if that index is even, you must surround the variable with absolute value symbols ($||$) when you place it in front of the radical.

Simplifying Radicals With Numbers

$\sqrt{8}$
$\sqrt{2 \times 2 \times 2}$
$2\sqrt{2}$

$\sqrt{50}$
$\sqrt{2 \times 5 \times 5}$
$5\sqrt{2}$

$\sqrt[3]{16}$
$\sqrt[3]{2 \times 2 \times 2 \times 2}$
$2\sqrt[3]{2}$

$\sqrt{27}$
$\sqrt{3 \times 3 \times 3}$
$3\sqrt{3}$

$\sqrt{300}$
$\sqrt{2 \times 2 \times 3 \times 5 \times 5}$
$2 \times 5\sqrt{3}$
$10\sqrt{3}$

$\sqrt[3]{54}$
$\sqrt[3]{2 \times 3 \times 3 \times 3}$
$3\sqrt[3]{2}$

---

1 **Determine which prime numbers you can multiply together to equal the number under the radical sign ($\sqrt{}$).**

Note: A prime number is a number that you can only evenly divide by itself and the number 1. For example, 2, 3, 5 and 7 are all prime numbers. For help with finding the prime numbers you multiply together, see the top of page 171.

2 **Replace the number under the radical sign ($\sqrt{}$) with the numbers you determined in step 1.**

3 **Determine if any numbers appear under the radical sign the number of times indicated by the index. If so, remove the repeating numbers from under the radical sign and place one of the numbers to the left of the radical sign.**

Note: The index is the small number that appears in front of a radical sign. If a small number does not appear in front of a radical sign, assume the number is 2.

4 **Repeat step 3 for each number that appears the number of times indicated by the index.**

5 **If you have placed more than one number in front of the radical sign, multiply the numbers together.**

Tip

Is there a quick way to find all the prime numbers that multiply together to equal a number?

Yes. First, find any two numbers that multiply together to equal the number. For example, 4 × 9 equals 36. Then find the prime numbers that multiply together to equal each of these numbers. For example, 2 × 2 equals 4 and 3 × 3 equals 9. So, in this case, 2 × 2 × 3 × 3 are the prime numbers that multiply together to equal 36.

Practice

$a + b = c$

Simplify the following radicals. You can check your answers on page 261.

1) $\sqrt{9}$

2) $\sqrt[3]{80}$

3) $\sqrt{200y^3}$

4) $\sqrt{75x^5}$

5) $\sqrt{35}$

6) $\sqrt[4]{32}$

Simplifying Radicals With Variables

$\sqrt{x^5}$
$\sqrt{x \times x \times x \times x \times x}$
$x \times x \sqrt{x}$
$x^2 \sqrt{x}$

$\sqrt{18x^2}$
$\sqrt{2 \times 3 \times 3 \times x \times x}$
$3 \times |x| \sqrt{2}$
$3|x|\sqrt{2}$

$\sqrt{y^7}$
$\sqrt{y \times y \times y \times y \times y \times y \times y}$
$y \times y \times y \sqrt{y}$
$y^3 \sqrt{y}$

$\sqrt{12x^5y^2}$
$\sqrt{2 \times 2 \times 3 \times x \times x \times x \times x \times x \times y \times y}$
$2 \times x \times x \times |y| \sqrt{3 \times x}$
$2x^2 |y| \sqrt{3x}$

$\sqrt[3]{x^4y^6}$
$\sqrt[3]{x \times x \times x \times x \times x \times y \times y \times y \times y \times y \times y}$
$x \times y \times y \sqrt[3]{x}$
$xy^2 \sqrt[3]{x}$

$\sqrt[3]{40a^7}$
$\sqrt[3]{2 \times 2 \times 2 \times 5 \times a \times a}$
$2 \sqrt[3]{5 \times a \times a}$
$2 \sqrt[3]{5a^2}$

1. Replace each variable under the radical sign with the number of variables that multiply together to equal the variable's exponent.

2. Determine if a variable appears under the radical sign the number of times indicated by the index. If so, remove the repeating variables from under the radical sign and place one of the variables to the left of the radical sign.

3. Repeat step 2 for each variable that appears the number of times indicated by the index.

4. If you have placed more than one variable in front of the radical sign, multiply the variables together.

- If a number also appears under the radical sign, use the steps you learned on page 170 to simplify the numbers under the radical sign.

- If the index of a radical is an even number and you have placed an instance of a variable to the left of a radical sign, surround the variable with absolute value symbols (||). This ensures that the answer will be positive even if the variable is negative.

Note: For more information on absolute values, see page 24.

Add and Subtract Radicals

Now that working with variables is second nature for you, adding and subtracting radicals will be a piece of cake—the process is virtually identical to adding and subtracting variables. Just as you can only add and subtract similar variables, you can only add and subtract similar, or like, radicals. The radicals must have the same index, which is the small number that appears before the radical sign, and radicand, which is the number and/or variable, or term, under the radical sign.

Before doing any adding or subtracting, you must check the index and radicand of each radical you are working with. If the radicals have the same index and radicand, the process is simple. You just add or subtract the coefficients, which are the numbers in front of the radicals. The radicals themselves remain unaltered. For example, $7\sqrt{x} - 4\sqrt{x}$ equals $3\sqrt{x}$.

Radicals with a different index and/or radicand are like oil and water when it comes to addition and subtraction—they do not mix. For example, $2\sqrt{a} + 5\sqrt{b}$ does not equal $7\sqrt{ab}$.

Add Radicals

Index

$$3\sqrt[3]{2} + 6\sqrt[3]{2} = 9\sqrt[3]{2}$$

$$5\sqrt{3} + \sqrt{3} = 6\sqrt{3}$$

$$2\sqrt{5} + 4\sqrt{5} + \sqrt{5} = 7\sqrt{5}$$

$$6\sqrt[3]{7} + 8\sqrt[3]{7} = 14\sqrt[3]{7}$$

$$2\sqrt[3]{4} + 3\sqrt[3]{4} = 5\sqrt[3]{4}$$

$$3\sqrt{x} + 6\sqrt{x} = 9\sqrt{x}$$

$$10\sqrt{y} + \sqrt{y} = 11\sqrt{y}$$

$$2\sqrt{xy} + 9\sqrt{xy} + 5\sqrt{xy} = 16\sqrt{xy}$$

$$7\sqrt[3]{2x^2} + 4\sqrt[3]{2x^2} = 11\sqrt[3]{2x^2}$$

$$5\sqrt[3]{a} + 8\sqrt[3]{a} = 13\sqrt[3]{a}$$

- You can only add radicals if the indexes as well as the numbers under the radical signs ($\sqrt{\ }$) are the same.

  Note: The index is the small number before the radical sign, such as 3 in $\sqrt[3]{2}$. If a small number does not appear before a radical sign, assume the number is 2.

1 To add radicals, add the numbers in front of the radical signs and place the result in front of a radical sign. Keep the number under the radical sign the same.

  Note: If a number does not appear in front of a radical sign, assume the number is 1. For example, $\sqrt{3}$ equals $1\sqrt{3}$.

- You can also add radicals that have a variable under the radical sign.

- As with numbers under the radical sign, if radicals have a variable under the radical sign, you can only add the radicals if the indexes as well as the variables and their exponents are the same.

Tip

Can I add or subtract radicals that have different radicands, or numbers and/or variables, under the radical signs?

Yes, but only if you can simplify one or both radicals to make the radicands the same. When the radicands are the same, you can add or subtract the radicals. In the following example, you can simplify $5\sqrt{12}$ to $10\sqrt{3}$ so that you can add this radical to the radical $2\sqrt{3}$. For information on simplifying radicals, see page 170.

$$5\sqrt{12} + 2\sqrt{3}$$
$$= 5(2\sqrt{3}) + 2\sqrt{3}$$
$$= 10\sqrt{3} + 2\sqrt{3}$$
$$= 12\sqrt{3}$$

Practice

Add or subtract the following radical expressions. You can check your answers on page 261.

1) $3\sqrt{2} - 7\sqrt{2}$
2) $5\sqrt[3]{10} + 6\sqrt[3]{10}$
3) $\sqrt{75} + 4\sqrt{3}$
4) $\sqrt{12} + \sqrt{300}$
5) $\sqrt[3]{9} + 4\sqrt[3]{9}$
6) $\sqrt[3]{54} + \sqrt[3]{2}$

Subtract Radicals

Index

$$7\sqrt[3]{3} - 5\sqrt[3]{3} = 2\sqrt[3]{3}$$

$$15\sqrt{6} - \sqrt{6} = 14\sqrt{6}$$

$$10\sqrt{7} - 2\sqrt{7} - 4\sqrt{7} = 4\sqrt{7}$$

$$6\sqrt[3]{5} \quad 5\sqrt[4]{5} - \sqrt[3]{5}$$

$$9\sqrt[3]{4} - 5\sqrt[3]{4} = 4\sqrt[3]{4}$$

$$12\sqrt{x} - 8\sqrt{x} = 4\sqrt{x}$$

$$16\sqrt{a} - \sqrt{a} = 15\sqrt{a}$$

$$8\sqrt{ab} - 3\sqrt{ab} - 2\sqrt{ab} = 3\sqrt{ab}$$

$$20\sqrt[3]{5x} - 10\sqrt[3]{5x} = 10\sqrt[3]{5x}$$

$$30\sqrt[3]{y^2} - 15\sqrt[3]{y^2} = 15\sqrt[3]{y^2}$$

- You can only subtract radicals if the indexes as well as the numbers under the radical signs ($\sqrt{\ }$) are the same.

Note: The index is the small number before the radical sign, such as 3 in $\sqrt[3]{2}$. If a small number does not appear before a radical sign, assume the number is 2.

1 To subtract radicals, subtract the numbers in front of the radical signs and place the result in front of a radical sign. Keep the number under the radical sign the same.

- You can also subtract radicals that have a variable under the radical sign.

- As with numbers under the radical sign, if radicals have a variable under the radical sign, you can only subtract the radicals if the indexes as well as the variables are the same.

Multiply Radicals

Multiplying radicals is pretty straightforward. Unlike adding and subtracting radicals, the radicals do not need to have the same number and/or variable, called the radicand, under the radical sign. The radicals must, however, have the same index, which is the small number just in front of the radical sign.

When multiplying two radicals, you first multiply any coefficients, or terms in front of the radicals, and then multiply the terms under the radical signs. The answer will have one radical sign with an index that matches the indexes of the original radicals. You place the product of the coefficients to the left of the radical sign and place the product of the radicands under the radical sign. For example, $4\sqrt{5} \times 3\sqrt{6}$ equals $12\sqrt{30}$.

After multiplying radicals, you should check whether the resulting radical can be simplified. For example, the radical $\sqrt{20}$ can be broken down to $\sqrt{(4)(5)}$. Since 4 is equal to 2^2, you can remove 2^2 from beneath the radical sign to give you a simplified expression of $2\sqrt{5}$.

Multiply Radicals With Numbers

Index

$$2\sqrt[3]{7} \times 8\sqrt[3]{3} = 16\sqrt[3]{21}$$

$$3\sqrt{2} \times 6\sqrt{8} = 18\sqrt{16}$$
$$= 18\sqrt{2 \times 2 \times 2 \times 2}$$
$$= 18 \times 2 \times 2$$
$$= 72$$

$$4\sqrt{6} \times \sqrt{6} = 4\sqrt{36}$$
$$= 4\sqrt{2 \times 2 \times 3 \times 3}$$
$$= 4 \times 2 \times 3$$
$$= 24$$

$$7\sqrt{3} \times 3\sqrt{9} = 21\sqrt{27}$$
$$= 21\sqrt{3 \times 3 \times 3}$$
$$= 21 \times 3\sqrt{3}$$
$$= 63\sqrt{3}$$

$$9\sqrt[3]{10} \times 3\sqrt[3]{4} = 27\sqrt[3]{40}$$
$$= 27\sqrt[3]{2 \times 2 \times 2 \times 5}$$
$$= 27 \times 2\sqrt[3]{5}$$
$$= 54\sqrt[3]{5}$$

$$5\sqrt[3]{2} \times 2\sqrt[3]{9} = 10\sqrt[3]{18}$$

- You can only multiply radicals if both radicals have the same index.

  Note: The index is the small number before the radical sign, such as the 3 in $\sqrt[3]{7}$. If a small number does not appear before a radical sign, assume the index is 2.

1 To multiply radicals, first multiply the numbers in front of the radical signs and write the result in front of a new radical sign, using the same index.

  Note: If a number does not appear in front of a radical sign, assume the number is 1. For example, $\sqrt{6}$ equals $1\sqrt{6}$.

2 Multiply the numbers under the radical signs and write the result under the new radical sign.

3 If your answer can be simplified, simplify the answer.

  Note: For information on simplifying radicals, see page 170.

Practice

$a + b = c$

Multiply the following radical expressions, making sure you simplify your answers. You can check your answers on page 261.

1) $\sqrt{2} \times \sqrt{7}$

2) $\sqrt{11} \times \sqrt{11}$

3) $\sqrt[3]{4} \times \sqrt[3]{6}$

4) $\sqrt{10} \times \sqrt{35}$

5) $\sqrt[3]{4} \times \sqrt[3]{5}$

6) $\sqrt[4]{100} \times \sqrt[4]{200}$

Multiply Radicals With Variables

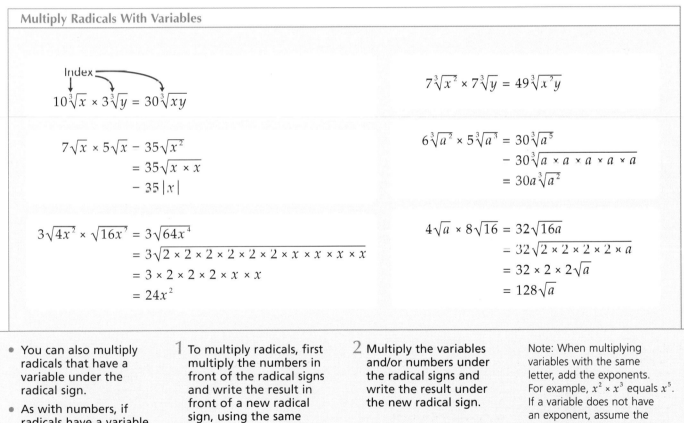

Index

$10\sqrt[3]{x} \times 3\sqrt[3]{y} = 30\sqrt[3]{xy}$

$7\sqrt[3]{x^2} \times 7\sqrt[3]{y} = 49\sqrt[3]{x^2 y}$

$$7\sqrt{x} \times 5\sqrt{x} - 35\sqrt{x^2}$$
$$= 35\sqrt{x \times x}$$
$$- 35\,|x|$$

$$6\sqrt[3]{a^2} \times 5\sqrt[3]{a^3} = 30\sqrt[3]{a^5}$$
$$- 30\sqrt[3]{a \times a \times a \times a \times a}$$
$$= 30a\sqrt[3]{a^2}$$

$$3\sqrt{4x^2} \times \sqrt{16x^2} = 3\sqrt{64x^4}$$
$$= 3\sqrt{2 \times 2 \times 2 \times 2 \times 2 \times 2 \times x \times x \times x \times x}$$
$$= 3 \times 2 \times 2 \times 2 \times x \times x$$
$$= 24x^2$$

$$4\sqrt{a} \times 8\sqrt{16} = 32\sqrt{16a}$$
$$= 32\sqrt{2 \times 2 \times 2 \times 2 \times a}$$
$$= 32 \times 2 \times 2\sqrt{a}$$
$$= 128\sqrt{a}$$

- You can also multiply radicals that have a variable under the radical sign.

- As with numbers, if radicals have a variable under the radical sign ($\sqrt{}$), you can only multiply the radicals if both radicals have the same index.

1 To multiply radicals, first multiply the numbers in front of the radical signs and write the result in front of a new radical sign, using the same index.

2 Multiply the variables and/or numbers under the radical signs and write the result under the new radical sign.

Note: When multiplying variables with the same letter, add the exponents. For example, $x^2 \times x^3$ equals x^5. If a variable does not have an exponent, assume the exponent is one. For example, x is the same as x^1.

3 If your answer can be simplified, simplify the answer.

Divide Radicals

To divide two radicals, the index, which is the small number just in front of the radical sign ($\sqrt{}$), of each radical must be the same. The radicals do not, however, need to have the same radicand, which is the number and/or variable under the radical sign.

When you divide radicals, the coefficients, or values in front of the radical, and the radicands are divided separately. In the answer, the result of dividing the coefficients is placed to the left of a new radical sign, while the result of dividing the radicands is placed underneath the radical sign. If the radicand cannot be divided evenly, the resulting fraction is split up, giving the numerator, or top part of the fraction, and the denominator, or bottom part of the fraction, their own radical signs. For example, $\sqrt{\frac{3}{2}}$ becomes $\frac{\sqrt{3}}{\sqrt{2}}$.

If the original radical had an index of 2, you can multiply both the numerator and denominator by the value of the denominator, effectively removing the radical sign from the denominator—a process called rationalization. Rationalization is expected by some teachers, who insist that it is bad math manners to have a radical in the denominator of a fraction.

Divide Radicals

Index

$$24\sqrt[3]{14} \div 6\sqrt[3]{2} = 4\sqrt[3]{7}$$

$$12\sqrt{27} \div 4\sqrt{3} = 3\sqrt{9}$$
$$= 3\sqrt{3 \times 3}$$
$$= 3 \times 3$$
$$= 9$$

$$9\sqrt{x^6} \div \sqrt{x^2} = 9\sqrt{x^{6-2}}$$
$$= 9\sqrt{x^4}$$
$$= 9\sqrt{x \times x \times x \times x}$$
$$= 9 \times x \times x$$
$$= 9x^2$$

$$40\sqrt{20x^5} \div 8\sqrt{5x^4} = 5\sqrt{4x^{5-4}}$$
$$= 5\sqrt{4x}$$
$$= 5\sqrt{2 \times 2 \times x}$$
$$= 5 \times 2\sqrt{x}$$
$$= 10\sqrt{x}$$

$$100\sqrt[3]{30} \div 10\sqrt[3]{2} = 10\sqrt[3]{15}$$

$$15\sqrt[3]{x^4y^6} \div 3\sqrt[3]{x^2y^5} = 5\sqrt[3]{x^{4-2}y^{6-5}}$$
$$= 5\sqrt[3]{x^2y}$$

• You can only divide radicals if both radicals have the same index.

Note: The index is the small number before the radical sign, such as 3 in $\sqrt[3]{14}$. If a small number does not appear before a radical sign, assume the index is 2.

1 To divide radicals, first divide the numbers in front of the radical signs and write the result in front of a new radical sign, using the same index.

Note: If a number does not appear in front of a radical sign, assume the number is 1. For example, $\sqrt{x}$ equals $1\sqrt{x}$.

2 Divide the numbers and/or variables under the radical signs and write the result under the new radical sign.

Note: When dividing variables with the same letter, subtract the exponents. For example, $x^6 \div x^2$ equals x^4. If a variable does not have an exponent, assume the exponent is one. For example, x is the same as x^1.

3 If your answer can be simplified, simplify the answer.

Note: For information on simplifying radicals, see page 170.

Tip

What should I do if the numbers in front of the radicals do not divide evenly?

You can write the numbers as a fraction in front of the radical sign in your answer. If you can simplify the fraction by dividing both the top and bottom part of the fraction by the same number, you should do so. For more information on simplifying fractions, see page 45. The following is an example of dividing radicals when the numbers in front of the radicals do not evenly divide.

$$10\sqrt{18} \div 4\sqrt{6}$$

$$= \frac{10\sqrt{3}}{4}$$

$$= \frac{5\sqrt{3}}{2}$$

Practice

$a+b=c$

Divide the following radical expressions. Make sure you simplify your answers and rationalize the denominator when you can. You can check your answers on page 261.

1) $6\sqrt{14} \div 3\sqrt{7}$

2) $9\sqrt{20x} \div 3\sqrt{5}$

3) $10\sqrt[3]{20x^2} \div 2\sqrt[3]{5x}$

4) $\sqrt{11} \div \sqrt{5}$

5) $36\sqrt{7} \div 12\sqrt{3}$

6) $7\sqrt{x^5 y^4} \div 3\sqrt{x^3 y^2}$

Rationalize the Denominator

$$80\sqrt{3} : 10\sqrt{2} = \frac{8\sqrt{3}}{\sqrt{2}}$$

$$= \frac{8\sqrt{3}}{\sqrt{2}} \times \frac{\sqrt{2}}{\sqrt{2}}$$

$$= \frac{8\sqrt{6}}{2}$$

$$= 4\sqrt{6}$$

$$45\sqrt{x} : 5\sqrt{y} = \frac{9\sqrt{x}}{\sqrt{y}}$$

$$= \frac{9\sqrt{x}}{\sqrt{y}} \times \frac{\sqrt{y}}{\sqrt{y}}$$

$$= \frac{9\sqrt{xy}}{y}$$

$$\sqrt{12x^2} \div \sqrt{5x} = \frac{\sqrt{12x^{2-1}}}{\sqrt{5}}$$

$$= \frac{\sqrt{12x}}{\sqrt{5}} \times \frac{\sqrt{5}}{\sqrt{5}}$$

$$= \frac{\sqrt{60x}}{5}$$

$$= \frac{\sqrt{2 \times 2 \times 3 \times 5 \times x}}{5}$$

$$= \frac{2\sqrt{15x}}{5}$$

1 When dividing radicals, if the numbers and/or variables under the radical signs do not divide evenly, write the radicals as a fraction, with each part of the fraction under its own radical sign.

• When your answer contains a radical in the denominator, you should remove the radical from the denominator to simplify your answer. This is known as rationalizing the denominator.

Note: The denominator is the bottom part of a fraction.

2 To remove a radical with an index of 2 from the denominator in your answer, multiply the top and bottom part of the fraction by the radical in the denominator.

Note: For information on how to multiply radicals, see page 174.

3 If your answer can be simplified, simplify the answer.

Write Radicals as Exponents

Instead of using a radical sign ($\sqrt{}$), a radical can also be expressed as a number or variable with a fraction as an exponent. For example, $\sqrt[3]{4^2}$ can be expressed as $4^{\frac{2}{3}}$ and read as "four to the power of two-thirds." Just when you were thinking that exponents were hard enough to read, now you have tiny fractional exponents to deal with.

When converting a radical expression to an exponential expression, the exponent of the radicand, or term under the radical sign, becomes the top number, or numerator, of the fractional exponent. The index, or small number in front of the radical sign, becomes the bottom number, or denominator, of the fractional exponent. For example, $\sqrt[y]{a^x}$ would convert to $a^{\frac{x}{y}}$.

Conversely, you can also convert expressions with fractional exponents to radical expressions. In this case, the top number of the exponential fraction becomes the exponent of the radicand, while the bottom number of the fraction becomes the index of the new radical.

Change a Radical Expression to an Exponential Expression

$$\sqrt[4]{5^3} = 5^{\frac{3}{4}} \qquad\qquad \sqrt[3]{x^2} = x^{\frac{2}{3}}$$

$$\sqrt[3]{4^2} = 4^{\frac{2}{3}} \qquad\qquad \sqrt[3]{(xy)^4} = (xy)^{\frac{4}{3}}$$

$$\sqrt{7} = 7^{\frac{1}{2}} \qquad\qquad \sqrt[3]{(a^3b^4)^2} = (a^3b^4)^{\frac{2}{3}}$$

$$\sqrt{x} = x^{\frac{1}{2}} \qquad\qquad \sqrt[4]{xy} = (xy)^{\frac{1}{4}}$$

$$\sqrt[3]{x} = x^{\frac{1}{3}} \qquad\qquad \sqrt{2a} = (2a)^{\frac{1}{2}}$$

- You can write a radical expression without a radical sign ($\sqrt{}$). The new expression will have an exponent that is a fraction.

1 Write the number and/or variables under the radical sign without a radical sign.

2 If two or more numbers and/or variables appear under the radical sign, surround the numbers and/or variables with parentheses ( ).

3 Write the exponent of the numbers and/or variables that appear under the radical sign as the top part of the fractional exponent.

Note: If the numbers and/or variables that appear under the radical sign do not have an exponent, assume the exponent is 1.

4 Write the index of the radical as the bottom part of the fractional exponent.

Note: The index is the small number before the radical sign, such as 3 in $\sqrt[3]{x^2}$. If an index is not shown, assume the index is 2.

Tip

How do I determine which number should become the top part of the new fractional exponent?

When writing a radical expression without a radical sign, determining the number you will use as the top part of the fraction can be tricky. If parentheses ( ) do not appear around two or more numbers and/or variables under the radical sign, no matter what the exponents of the individual numbers and variables are, consider the exponents to be 1. Then use the number 1 as the top part of the fraction, as shown in the following examples.

$$\sqrt{3a^3} = (3a^3)^{\frac{1}{2}}$$

$$\sqrt[3]{x^2y^4} = (x^2y^4)^{\frac{1}{3}}$$

Practice

$a + b = c$

Change the following radical expressions to exponential expressions. You can check your answers on page 261.

1) $\sqrt[3]{2^2}$

2) $\sqrt[7]{3x^2}$

3) $\sqrt[4]{y^2}$

Change the following exponential expressions to radical expressions. You can check your answers on page 261.

4) $5^{\frac{3}{4}}$

5) $x^{\frac{2}{5}}$

6) $y^{\frac{1}{3}}$

Change an Exponential Expression to a Radical Expression

$$5^{\frac{2}{3}} = \sqrt[3]{5^2} \qquad\qquad x^{\frac{1}{2}} = \sqrt{x}$$

$$4^{\frac{3}{2}} = \sqrt{4^3} \qquad\qquad (7x)^{\frac{2}{5}} = \sqrt[5]{(7x)^2}$$

$$8^{\frac{1}{5}} = \sqrt[5]{8} \qquad\qquad (9xy)^{\frac{1}{3}} = \sqrt[3]{(9xy)} = \sqrt[3]{9xy}$$

$$11^{\frac{4}{3}} = \sqrt[3]{11^4} \qquad\qquad 6x^{\frac{3}{5}} = 6\sqrt[5]{x^3}$$

$$x^{\frac{2}{3}} = \sqrt[3]{x^2} \qquad\qquad 14x^{\frac{2}{3}} = 14\sqrt[3]{x^2}$$

- You can write an expression with a fractional exponent as a radical expression.

1 Write the number or variable to the left of the exponent under a radical sign ($\sqrt{}$).

Note: If parentheses appear around a number and/or variables to the left of an exponent, place everything in the parentheses, including the parentheses, under the radical sign. Otherwise, place only the number or variable directly to the left of the exponent under the radical sign.

2 Write the number in the top part of the fraction as the exponent for the number and/or variables under the radical sign.

Note: If the number in the top part of the fraction is 1, you do not need to write an exponent for the number and/or variables under the radical sign.

3 Write the number in the bottom part of the fraction as the index for the radical.

Note: The index is the small number before the radical sign, such as 3 in $\sqrt[3]{5^2}$. When the index is 2, you do not need to write the index.

Solve a Radical Equation

Equations that contain radicals may look intimidating, but solving them is straightforward.

The one thing holding you back from being able to solve an equation that contains a radical expression is the radical expression itself, so your primary goal is to get rid of it. You will need to isolate the radical expression to one side of the equation the same way you would isolate a variable in any equation you want to solve. Once the radical expression is isolated, you can make the radical symbol disappear by raising the radical to the power of the index—

that little number just in front of the radical symbol. For example, you would raise $\sqrt[3]{x}$ to the 3rd power, leaving x standing by itself.

As is the case with any equation, balance is key—whatever you do to one side of the equation, you must do to the other side. So you must also raise the side of the equation without the radical expression to the same power to which you raised the side containing the radical expression. From that point, it is just a simple matter of solving for the variable.

Step 1: Isolate the Radical

$$2\sqrt{3x + 1} + 4 = 8$$

$$2\sqrt{3x + 1} + 4 - 4 = 8 - 4$$
$$2\sqrt{3x + 1} = 4$$
$$\frac{2\sqrt{3x + 1}}{2} = \frac{4}{2}$$
$$\sqrt{3x + 1} = 2$$

Step 2: Remove the Radical Sign

$$\sqrt{3x + 1} = 2$$
$$(\sqrt{3x + 1})^2 = 2^2$$
$$3x + 1 = 4$$

---

1 Determine which numbers you need to add, subtract, multiply and/or divide to place the radical by itself on one side of the equation, usually the left side.

2 Add, subtract, multiply and/or divide both sides of the equation by the numbers you determined in step 1.

• In this example, subtract 4 from both sides of the equation. Then divide both sides of the equation by 2.

Note: For more information on adding, subtracting, multiplying and dividing numbers in equations, see pages 64 to 67.

3 Determine the index of the radical.

Note: The index is the small number in front of the radical sign. If a small number does not appear in front of a radical sign, assume the index is 2.

4 To remove the radical sign ($\sqrt{}$), use an exponent to raise both sides of the equation to the same number as the index.

• In this example, since the index is 2, raise both sides of the equation to the power of 2.

Tip

What should I do if I end up with a negative variable?

If you end up with a negative sign (–) in front of a variable, such as –x = 10, you will have to make the variable positive to finish solving the equation. Fortunately, this is a simple task—just multiply both sides of the equation by –1. The equation remains balanced, while the variable switches from a negative to a positive value. In the case of –x = 10, you would end up with a solution of x = –10.

Practice

Solve for the variable in the following radical equations. You can check your answers on page 261.

1) $2\sqrt{x-1} = 10$

2) $\sqrt{2x+4} = 2$

3) $\sqrt[3]{x+3} = -3$

4) $\sqrt[4]{11-5x} = 1$

5) $\sqrt[3]{5-x} = -8$

6) $\sqrt{4x-7} = -2$

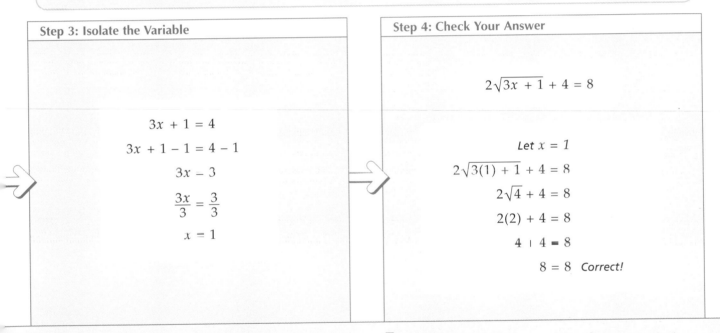

Step 3: Isolate the Variable

$$3x + 1 = 4$$
$$3x + 1 - 1 = 4 - 1$$
$$3x = 3$$
$$\frac{3x}{3} = \frac{3}{3}$$
$$x = 1$$

Step 4: Check Your Answer

$$2\sqrt{3x+1} + 4 = 8$$

Let x = 1
$$2\sqrt{3(1)+1} + 4 = 8$$
$$2\sqrt{4} + 4 = 8$$
$$2(2) + 4 = 8$$
$$4 + 4 = 8$$
$$8 = 8 \quad \textit{Correct!}$$

5 Determine which numbers you need to add, subtract, multiply and/or divide to place the variable on one side of the equation, usually the left side.

6 Add, subtract, multiply and/or divide both sides of the equation by the numbers you determined in step 5.

- In this example, subtract 1 from both sides of the equation. Then divide both sides of the equation by 3.

7 To check your answer, place the number you found into the original equation and solve the problem. If both sides of the equation are equal, you have correctly solved the equation.

Note: In some cases, both sides of the equation will not end up being equal when you check your answer. This can occur when an equation has no solution, as in the equation $\sqrt{x} = -2$. After you make sure you did not make an error when solving the radical equation, you can discard the answer as a false solution.

Working With Radicals

Question 1. Simplify the following radicals as much as possible.

a) $\sqrt{4}$

b) $\sqrt{50}$

c) $\sqrt[3]{27}$

d) $\sqrt[4]{2^5}$

e) $\sqrt{48}$

f) $\sqrt{x^2}$

g) $\sqrt[3]{y^3}$

h) $\sqrt{z^3}$

Question 2. Add or subtract the following radical expressions.

a) $\sqrt{2} + 5\sqrt{2}$

b) $\sqrt[3]{5} + 4\sqrt[3]{5}$

c) $5\sqrt{2} + \sqrt{3} - 3\sqrt{2} + 9\sqrt{3}$

d) $\sqrt{3} + \sqrt{12}$

e) $\sqrt[3]{5} + 3\sqrt{5} - 4\sqrt[3]{5} + 7\sqrt{5}$

Question 3. Multiply the following radical expressions. Simplify your answers as much as possible.

a) $\sqrt{5} \times \sqrt{7}$

b) $\sqrt{2} \times \sqrt{2}$

c) $\sqrt[3]{4} \times \sqrt[3]{4}$

d) $\sqrt[5]{10} \times \sqrt[5]{8}$

e) $\sqrt[3]{x} \times \sqrt[3]{x} \times \sqrt[3]{x}$

Question 4. Divide the following radical expressions. Simplify your answers as much as possible.

a) $\sqrt{10} \div \sqrt{5}$

b) $\sqrt{21} \div \sqrt{7}$

c) $\sqrt[3]{100} \div \sqrt[3]{5}$

d) $\sqrt[5]{49} \div \sqrt[5]{7}$

e) $\sqrt{x^6} \div \sqrt{x^2}$

Question 5. Write the following radical expressions as exponential expressions.

a) $\sqrt{5}$

b) $\sqrt[3]{10}$

c) $\sqrt[5]{3^2}$

d) $\sqrt{x}$

e) $\sqrt[4]{y^7}$

Question 6. Write the following exponential expressions as radical expressions.

a) $6^{\frac{1}{2}}$

b) $7^{\frac{1}{2}}$

c) $z^{\frac{1}{3}}$

d) $x^{\frac{4}{5}}$

e) $3^{\frac{5}{4}}$

Question 7. Solve for the variable in the following equations.

a) $\sqrt{x} = 3$

b) $\sqrt{x} = -2$

c) $\sqrt{x} + 5 = 9 - \sqrt{x}$

d) $3\sqrt[3]{x} = 5 + 2\sqrt[3]{x}$

e) $1 + \sqrt[5]{x} = 9 - 7\sqrt[5]{x}$

You can check your answers on page 277.

Chapter 10

F actoring is the mathematical equivalent of rewinding. When you factor an expression, it is like starting with the answer and working your way back to the question. Chapter 10 reveals a number of handy techniques for factoring expressions.

Factoring Expressions

In this Chapter...

Factor Using the Greatest Common Factor

Factoring is like pressing the rewind button on an algebra problem. You start out with an expression and work backward so that you end up with a multiplication problem. For example, 12 can be factored out to (6)(2). As you can see, factoring is like unmultiplying.

There are a number of ways to factor, but the quickest and most commonly used factoring technique is the Greatest Common Factor (GCF) method. The greatest common factor is the largest term that divides evenly into each term in an expression. When you factor using the greatest common factor method, you find and remove the greatest common factor, placing the rest of the expression in parentheses. For example, 3 is the greatest common factor of the expression 3 + 6 + 9, so the expression would be factored out to 3(1 + 2 + 3).

To check your answer, work out the expression using the distributive property. If you end up with the original expression, you have correctly factored the expression.

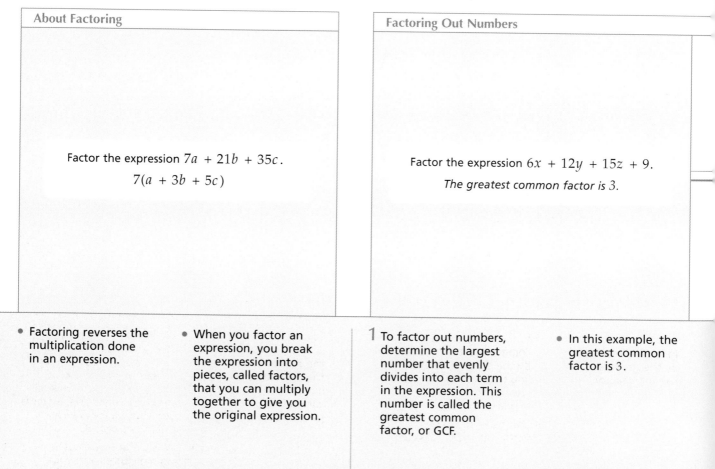

About Factoring

Factor the expression $7a + 21b + 35c$.

$7(a + 3b + 5c)$

Factoring Out Numbers

Factor the expression $6x + 12y + 15z + 9$.

The greatest common factor is 3.

- Factoring reverses the multiplication done in an expression.

- When you factor an expression, you break the expression into pieces, called factors, that you can multiply together to give you the original expression.

1 To factor out numbers, determine the largest number that evenly divides into each term in the expression. This number is called the greatest common factor, or GCF.

- In this example, the greatest common factor is 3.

Tip

How can I determine what is the greatest common factor of the numbers in an expression?

Begin by writing down all the numbers, called factors, that evenly divide into each number in the expression. The largest factor that all the numbers in the expression have in common is the greatest common factor. For instance, in the expression $24x + 30y$, write down all the numbers that evenly divide into 24 and 30. In this case, the greatest common factor is 6.

$$24x + 30y$$

Factors of 24 are: 1, 2, 3, 4, 6, 8, 12, 24
Factors of 30 are: 1, 2, 3, 5, 6, 10, 15, 30

Practice

Factor the following expressions using the greatest common factor method. You can check your answers on page 262.

1) $2 + 4 + 6$

2) $6 + 9 + 15$

3) $12 + 18 + 60$

4) $4y + 2x + 50z$

5) $10a + 100b + 50c$

6) $7a + 7b + 7$

$$6x + 12y + 15z + 9$$

$$\frac{6x}{3} + \frac{12y}{3} + \frac{15z}{3} + \frac{9}{3}$$

$$2x + 4y + 5z + 3$$

$$3(2x + 4y + 5z + 3)$$

$$3(2x + 4y + 5z + 3)$$
$$= 6x + 12y + 15z + 9 \quad \textit{Correct!}$$

---

2 Divide each term in the expression by the greatest common factor.

3 Write the greatest common factor followed by the division result you determined in step 2, surrounded by parentheses ( ).

- You have finished factoring the expression.

4 To check your answer, multiply the number outside the parentheses by each number and variable inside the parentheses.

Note: The distributive property allows you to remove a set of parentheses by multiplying each number and variable inside the parentheses by a number directly outside the parentheses. For more information on the distributive property, see page 30.

- If you end up with the original expression, you have correctly factored the expression.

Factor Using the Greatest Common Factor *continued*

Finding the Greatest Common Factor (GCF) in an expression requires that you find the largest term that divides evenly into each term in the expression.

The greatest common factor method is useful when factoring expressions that contain variables with exponents.

In an expression that contains variables with exponents, you need to look for variables that are common to each term in the expression and then choose the lowest exponent of those variables.

Then write the rest of the factored expression within parentheses. For example, in the expression $4x^4 + 5x^2 + 3x^6$, the greatest common factor is x^2. The factored expression would be written as $x^2(4x^2 + 5 + 3x^4)$.

To check your answer, expand the new factored expression. If the answer is the same as the original non-factored expression, you have solved correctly.

Factoring Out Variables

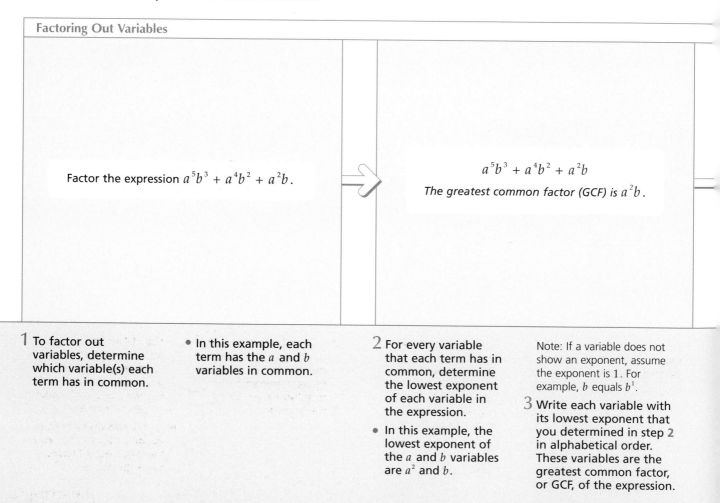

Factor the expression $a^5b^3 + a^4b^2 + a^2b$.

$$a^5b^3 + a^4b^2 + a^2b$$

The greatest common factor (GCF) is a^2b.

1 To factor out variables, determine which variable(s) each term has in common.

- In this example, each term has the a and b variables in common.

2 For every variable that each term has in common, determine the lowest exponent of each variable in the expression.

- In this example, the lowest exponent of the a and b variables are a^2 and b.

Note: If a variable does not show an exponent, assume the exponent is 1. For example, b equals b^1.

3 Write each variable with its lowest exponent that you determined in step **2** in alphabetical order. These variables are the greatest common factor, or GCF, of the expression.

Tip

Can I factor a negative variable out of an expression?

Yes. You can factor out a negative variable, such as $-a$, the same way you would factor out a positive variable, such as a. For example, to factor the expression $-a^5 - a^2$, you can factor out $-a^2$ by using $-a^2$ as the greatest common factor. Notice how the sign (+ or −) in front of each term changes.

$$-a^5 - a^2$$
$$\frac{-a^5}{-a^2} - \frac{a^2}{-a^2}$$
$$a^3 + a^0$$
$$a^3 + 1$$
$$-a^2(a^3 + 1)$$

Practice

Factor the following expressions using the greatest common factor method. You can check your answers on page 262.

1) $abc + acd + bce$
2) $x^2 + x^3 + x^{10}$
3) $x^2y^3z + x^4y^2z^2 + x^3y^4$
4) $a + a^3 - a^5$
5) $abc^3 - bc + b^4$
6) $-b^2 - b^4 + b^7$

$$a^5b^3 + a^4b^2 + a^2b$$
$$\frac{a^5b^3}{a^2b} + \frac{a^4b^2}{a^2b} + \frac{a^2b}{a^2b}$$
$$a^3b^2 + a^2b^1 + a^0b^0$$
$$a^3b^2 + a^2b + 1$$
$$a^2b(a^3b^2 + a^2b + 1)$$

$$a^2b(a^3b^2 + a^2b + 1)$$
$$= a^5b^3 + a^4b^2 + a^2b \quad \text{Correct!}$$

4 Divide each term in the expression by the greatest common factor.

Note: When you divide variables with exponents that have the same letter, you can subtract the exponents. For example, $a^5 \div a^2$ equals a^{5-2}, which equals a^3. A variable with the 0 exponent equals 1. For example, a^0 equals 1.

5 Write the greatest common factor followed by the division result you determined in step **4**, surrounded by parentheses ( ).

• You have finished factoring the expression.

6 To check your answer, multiply the number outside the parentheses by each variable and number inside the parentheses.

Note: When you multiply variables with exponents that have the same letter, you can add the exponents. For example, $a^2 \times a^3$ equals a^{2+3}, which equals a^5.

• If you end up with the original expression, you have correctly factored the expression.

CONTINUED

Factor Using the Greatest Common Factor *continued*

Factoring is a lot like a game show where contestants are given an answer and have to come up with the question. In factoring, you are given an expression and you must work out the original expression.

The Greatest Common Factor (GCF) method can be used to factor expressions that contain both numbers and variables that have exponents. First, determine the largest factor, or term that divides evenly into all the other terms, that each of the numbers in the expression has in common. Then find the lowest exponent of any variable that appears in each term of the expression. Next, place the number and variable with its exponent together to obtain the greatest common factor for the expression.

For example, if you have the expression $3a^4 + 6a^2 + 9a^3$, the largest common factor of the numbers in front of the variables, called coefficients, would be 3 and the variable with the lowest exponent would be a^2. Combining the two terms together, you arrive at a greatest common factor of $3a^2$.

As always, be sure to check your answer by expanding out the factored expression.

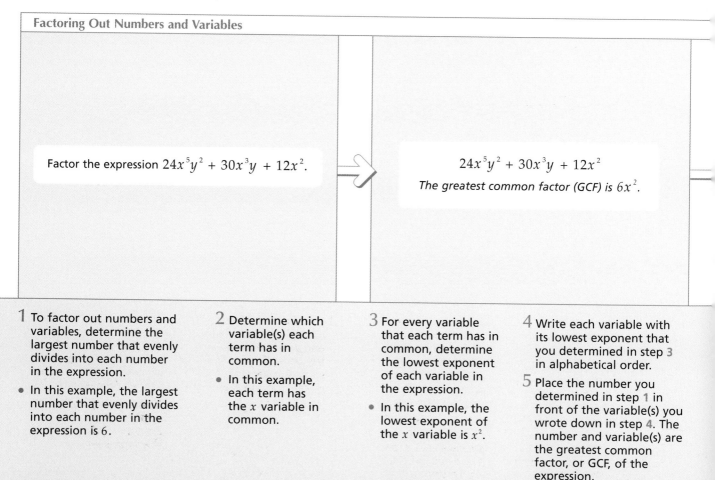

Factoring Out Numbers and Variables

Factor the expression $24x^5y^2 + 30x^3y + 12x^2$.

$24x^5y^2 + 30x^3y + 12x^2$

The greatest common factor (GCF) is $6x^2$.

1 To factor out numbers and variables, determine the largest number that evenly divides into each number in the expression.

- In this example, the largest number that evenly divides into each number in the expression is 6.

2 Determine which variable(s) each term has in common.

- In this example, each term has the x variable in common.

3 For every variable that each term has in common, determine the lowest exponent of each variable in the expression.

- In this example, the lowest exponent of the x variable is x^2.

4 Write each variable with its lowest exponent that you determined in step 3 in alphabetical order.

5 Place the number you determined in step 1 in front of the variable(s) you wrote down in step 4. The number and variable(s) are the greatest common factor, or GCF, of the expression.

Tip

Is there a common mistake I should watch out for?

One common mistake that people often make when factoring is to forget that a term divided by itself equals 1, not 0. If a term in an expression is exactly the same as the greatest common factor, when you divide the term by the greatest common factor, you are left with a value of 1. For example, in the expression $6x^3 + 9x^2 + 3x$, the term $3x$ is divided by the greatest common factor of $3x$, so make sure you place a 1 inside the parentheses, not a 0.

$$6x^3 + 9x^2 + 3x$$
$$= 3x(2x^2 + 3x + 1)$$

Practice

$a + b = c$

Factor the following expressions using the greatest common factor method. You can check your answers on page 262.

1) $2x + 4x^2 + 6x^3$

2) $10xy + 25x^2 + 35xy^2$

3) $8z^3 - 16z^5 + 20z^6$

4) $18xyz^2 - 81x^2y^3z$

5) $-4a^2 - 10a^5 + 6a - 8$

6) $9z^3 + 7x^2$

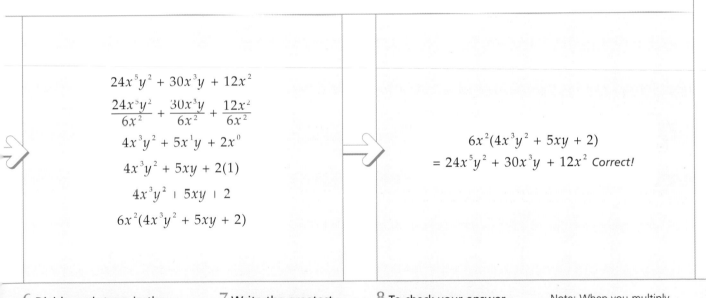

$$24x^5y^2 + 30x^3y + 12x^2$$

$$\frac{24x^5y^2}{6x^2} + \frac{30x^3y}{6x^2} + \frac{12x^2}{6x^2}$$

$$4x^3y^2 + 5x^1y + 2x^0$$

$$4x^3y^2 + 5xy + 2(1)$$

$$4x^3y^2 + 5xy + 2$$

$$6x^2(4x^3y^2 + 5xy + 2)$$

$$6x^2(4x^3y^2 + 5xy + 2)$$
$$= 24x^5y^2 + 30x^3y + 12x^2 \; Correct!$$

6 **Divide each term in the expression by the greatest common factor.**

Note: When you divide variables with exponents that have the same letter, you can subtract the exponents. For example, $x^5 \div x^2$ equals x^{5-2}, which equals x^3. A variable with the 0 exponent equals 1. For example, x^0 equals 1. A variable with the 1 exponent equals itself. For example, x^1 equals x.

7 **Write the greatest common factor followed by the division result you determined in step 6, surrounded by parentheses ( ).**

- You have finished factoring the greatest common factor out of the expression.

8 **To check your answer, multiply the number and variable outside the parentheses by each number and variable inside the parentheses.**

Note: When you multiply variables with exponents that have the same letter, you can add the exponents. For example, $x^2 \times x^3$ equals x^{2+3}, which equals x^5.

- If you end up with the original expression, you have correctly factored the expression.

Factor by Grouping

Factoring is like unmultiplying. When you factor an expression, you turn the expression back into a multiplication problem—showing the expression as the product of its smaller parts, which are called factors.

One specialized technique of factoring is the factor by grouping method. You cannot use this technique for every expression, but when you are able to factor by grouping, this method is fast. When an expression does not have any numbers or variables that are common to all of the terms, but does have some numbers and variables that are common to a few of the terms, you can try the grouping method of factoring.

To factor by grouping, you split the expression into two smaller groups of terms and then factor those groups individually. If the two factored parts of the original expression share a common factor, the groups can be reunited into one new factored expression. If the two groups do not turn out to have a factor in common, the expression cannot be factored by grouping.

While this method sounds complicated, being able to spot expressions that can be split into groups and factored will become second nature after a little practice.

Factor the expression $4ax + 10ay + 6bx + 15by$.

$$\overbrace{4ax + 10ay}^{\text{Group 1}} + \overbrace{6bx + 15by}^{\text{Group 2}}$$

$$4ax + 10ay = 2a(2x + 5y)$$
$$6bx + 15by = 3b(2x + 5y)$$

$$2a(2x + 5y) + 3b(2x + 5y)$$
The factor is $(2x + 5y)$.

- If all the terms in an expression do not have any numbers or variables in common, try to factor the expression by grouping.

1 Divide the terms into groups, with each group having numbers and/or variables in common. You will work with each group separately.

2 Using the greatest common factor method, factor out the numbers and/or variables that the terms in each group have in common.

Note: To use the greatest common factor method to factor out numbers and variables, see page 186.

3 Add the results of each group together.

4 Find the terms that each group of terms have in common. These terms are the factor for the new expression.

- In this example, each group of terms has the factor $(2x + 5y)$ in common.

Note: If the groups of terms do not have any factors in common, repeat steps 2 to 4, but try factoring out different numbers and/or variables in each group until the groups have factors in common.

Tip

When I factor by grouping, what should I do if the terms in parentheses are the same but have different signs (+ or –)?

After factoring an expression by grouping, you may need to factor out the negative sign so the terms in the parentheses exactly match. For example, in the expression $2a(2x + 5y) + 3b(-2x - 5y)$, you factor out the negative sign in the second set of parentheses to end up with $2a(2x + 5y) - 3b(2x + 5y)$. The terms in the parentheses now match, so you can finish factoring the expression.

Practice

Factor the following expressions using the grouping method. You can check your answers on page 262.

1) $ax + ay + bx + by$

2) $cx - dx + cy - dy$

3) $3ac - 2bd + 6bc - ad$

4) $2ax + 2bx + 3ay + 3by$

5) $5xz + 5x - yz - y$

6) $xy - 3x - 6 + 2y$

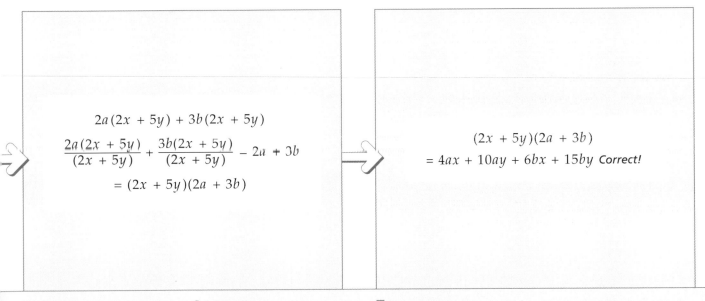

$$2a(2x + 5y) + 3b(2x + 5y)$$

$$\frac{2a(2x + 5y)}{(2x + 5y)} + \frac{3b(2x + 5y)}{(2x + 5y)} - 2a + 3b$$

$$= (2x + 5y)(2a + 3b)$$

$$(2x + 5y)(2a + 3b)$$

$$= 4ax + 10ay + 6bx + 15by \text{ Correct!}$$

5 Divide each group of terms by the factor you determined in step **4**.

Note: When dividing a group of terms such as $(2x + 5y)$, by the same group of terms, the groups of terms cancel each other out.

6 Write the factor followed by the division result you determined in step **5**, surrounded by parentheses ( ).

• You have finished factoring the expression.

7 To check your answer, you can multiply each term in the first set of parentheses by each term in the second set of parentheses.

Note: For more information on multiplying terms in parentheses, see page 154.

• If you end up with the original expression, even though the terms may be in a slightly different order, you have correctly factored the expression.

Recognizing Special Factor Patterns

Factoring is a process in which you show an expression as the product of its smaller parts, which are called factors.

In factoring, squares, which are the product of multiplying a value by itself, and cubes, which are the product of multiplying a value by itself three times, come in very handy.

You can save time when factoring expressions if you are able to recognize patterns that involve squares and cubes. For example, when you notice a pattern of one square being subtracted from another square, you can simply apply the difference of two squares formula, which states that $a^2 - b^2 = (a + b)(a - b)$. Similarly, if you see a pattern of two cubes being added together, you can simply apply the sum of two cubes formula, which states that $a^3 + b^3 = (a + b)(a^2 - ab + b^2)$.

To be able to recognize the patterns, you need to be able to recognize squares and cubes. Squares and cubes of variables are easy to recognize—terms that have a variable raised to an even-numbered exponent, such as a^2, a^4, a^6, are squares, while terms that have a variable raised to an exponent that is a multiple of three, such as a^3, a^6, a^9, are cubes.

The Difference of Two Squares

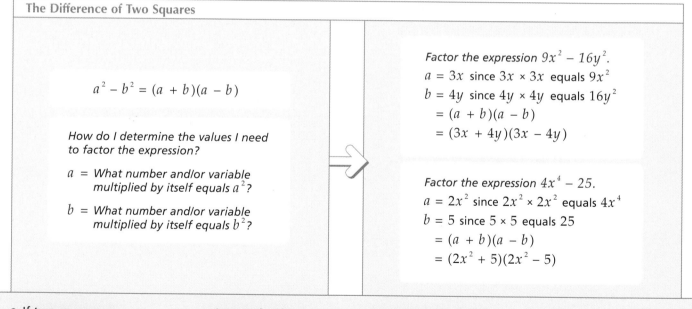

$$a^2 - b^2 = (a + b)(a - b)$$

How do I determine the values I need to factor the expression?

$a = $ *What number and/or variable multiplied by itself equals* a^2?

$b = $ *What number and/or variable multiplied by itself equals* b^2?

Factor the expression $9x^2 - 16y^2$.
$a = 3x$ since $3x \times 3x$ equals $9x^2$
$b = 4y$ since $4y \times 4y$ equals $16y^2$
$\quad = (a + b)(a - b)$
$\quad = (3x + 4y)(3x - 4y)$

Factor the expression $4x^4 - 25$.
$a = 2x^2$ since $2x^2 \times 2x^2$ equals $4x^4$
$b = 5$ since 5×5 equals 25
$\quad = (a + b)(a - b)$
$\quad = (2x^2 + 5)(2x^2 - 5)$

- If two squares are subtracted, such as $a^2 - b^2$, you can rewrite the expression as $(a + b)(a - b)$.
- A number or variable multiplied by itself is a square. For example, 3×3 equals 9 and $a \times a$ equals a^2.

- A term that has a variable raised to an even-numbered exponent, such as a^2, a^4, a^6, is a square. For example, a^4 equals $a^2 \times a^2$.

- In the above examples, you can use the difference of two squares method to factor the expressions.

Note: When you multiply variables with exponents that have the same letter, you add the exponents. For example, $x^2 \times x^2$ equals x^4.

Tip

What are the cubes of the numbers 1 to 10?

Here is a quick reference for the cubes of the numbers from 1 to 10. Once the numbers get higher, cubes can be more difficult to recognize. You should become familiar with the cubes of the numbers 1 through 10, since being able to recognize these numbers can save you time when working with algebra expressions.

$1^3 = 1$	$6^3 = 216$
$2^3 = 8$	$7^3 = 343$
$3^3 = 27$	$8^3 = 512$
$4^3 = 64$	$9^3 = 729$
$5^3 = 125$	$10^3 = 1,000$

Practice

$a + b = c$

Factor the following expressions. You can check your answers on page 262.

1) $x^2 - 9$
2) $y^4 - a^4$
3) $b^4 - 4$
4) $x^3 + y^3$
5) $a^3 + 8$
6) $b^6 + 27$

The Sum of Two Cubes

$$a^3 + b^3 = (a + b)(a^2 - ab + b^2)$$

How do I determine the values I need to factor the expression?

a = *What number and/or variable multiplied by itself three times equals a^3?*

b = *What number and/or variable multiplied by itself three times equals b^3?*

Factor the expression $x^3 + 125y^3$

$a = x$ since $x \times x \times x$ equals x^3

$b = 5y$ since $5y \times 5y \times 5y$ equals $125y^3$

$a^2 = x \times x = x^2$

$ab = x \times 5y = 5xy$

$b^2 = 5y \times 5y = 25y^2$

$\qquad = (a + b)(a^2 - ab + b^2)$

$\qquad = (x + 5y)(x^2 - 5xy + 25y^2)$

- If two cubes are added, such as $a^3 + b^3$, you can rewrite the expression as $(a + b)(a^2 - ab + b^2)$.

- A number or variable multiplied by itself three times is a cube. For example, $2 \times 2 \times 2$ equals 8 and $a \times a \times a$ equals a^3.

- A term that has a variable raised to an exponent that is a multiple of three, such as a^3, a^6, a^9, is a cube. For example, a^6 equals $a^2 \times a^2 \times a^2$.

- In the above example, you can use the sum of two cubes method to factor the expression.

CONTINUED

Recognizing Special Factor Patterns *continued*

Mathematicians have learned that squares, the product of a value multiplied by itself, and cubes, the product of a value multiplied by itself three times, follow specific patterns. These patterns are very useful for factoring.

In general, when you encounter an expression with two terms, you should check to see if both terms are perfect squares or cubes. If both terms are perfect squares or cubes, you can probably use one of the square or cube formulas to quickly factor the expression. In addition to the difference of two squares and the sum of two cubes patterns discussed on pages 194 and 195, you should also

be familiar with the pattern involving the difference of two cubes. In this instance, the formula $a^3 - b^3 = (a - b)(a^2 + ab + b^2)$ applies. If you encounter an expression in which one cube is subtracted from another, you can apply the difference of two cubes formula to easily factor out the expression.

Keep in mind that it is standard practice to factor an expression completely—in other words, factor until you can factor no more. After you factor an expression, you should check your answer to determine if you can factor the expression any further.

The Difference of Two Cubes

$$a^3 - b^3 = (a - b)(a^2 + ab + b^2)$$

How do I determine the values I need to factor the expression?

$a =$ *What number and/or variable multiplied by itself three times equals a^3?*

$b =$ *What number and/or variable multiplied by itself three times equals b^3?*

$\Rightarrow$

Factor the expression $8x^3 - 27y^3$

$a = 2x$ since $2x \times 2x \times 2x$ equals $8x^3$

$b = 3y$ since $3y \times 3y \times 3y$ equals $27y^3$

$a^2 = 2x \times 2x = 4x^2$

$ab = 2x \times 3y = 6xy$

$b^2 = 3y \times 3y = 9y^2$

$\quad = (a - b)(a^2 + ab + b^2)$

$\quad = (2x - 3y)(4x^2 + 6xy + 9y^2)$

- If two cubes are subtracted, such as $a^3 - b^3$, you can rewrite the expression as $(a - b)(a^2 + ab + b^2)$.

- A number or variable multiplied by itself three times is a cube. For example, $2 \times 2 \times 2$ equals 8 and $a \times a \times a$ equals a^3.

- A term that has a variable raised to an exponent that is a multiple of three, such as a^3, a^6, a^9, is a cube. For example, a^6 equals $a^2 \times a^2 \times a^2$.

- In the above example, you can use the difference of two cubes method to factor the expression.

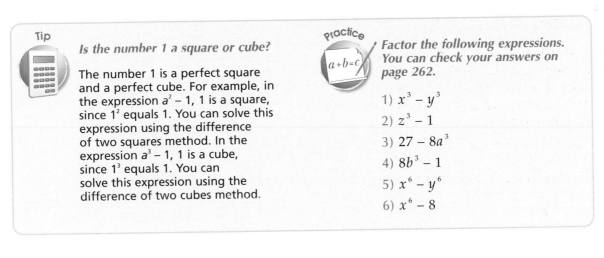

Tip

Is the number 1 a square or cube?

The number 1 is a perfect square and a perfect cube. For example, in the expression $a^2 - 1$, 1 is a square, since 1^2 equals 1. You can solve this expression using the difference of two squares method. In the expression $a^3 - 1$, 1 is a cube, since 1^3 equals 1. You can solve this expression using the difference of two cubes method.

Practice

Factor the following expressions. You can check your answers on page 262.

1) $x^3 - y^3$

2) $z^3 - 1$

3) $27 - 8a^3$

4) $8b^3 - 1$

5) $x^6 - y^6$

6) $x^6 - 8$

Factor the expression $x^6 - 64$

$a = x^2$ since $x^2 \times x^2 \times x^2$ equals x^6

$b = 4$ since $4 \times 4 \times 4$ equals 64

$a^2 = x^2 \times x^2 = x^4$

$ab = x^2 \times 4 = 4x^2$

$b^2 = 4 \times 4 = 16$

$= (a - b)(a^2 + ab + b^2)$

$= (x^2 - 4)(x^4 + 4x^2 + 16)$

Factor the expression $x^2 - 4$.

$a = x$ since $x \times x$ equals x^2

$b = 2$ since 2×2 equals 4

$= (a + b)(a - b)$

$= (x + 2)(x - 2)$

Final answer $= (x + 2)(x - 2)(x^4 + 4x^2 + 16)$

- In the above example, you can use the difference of two cubes method to factor the expression.

Note: When you multiply variables with exponents that have the same letter, you add the exponents. For example, $x^2 \times x^2$ equals x^4.

- In some cases, you will need to factor an expression more than once to reach your final answer. After you factor an expression, you should check your answer to determine if you can factor the expression further.

- In the above example, you can factor $(x^2 - 4)$ further.

- In the above example, you can use the difference of two squares method to finish factoring the expression.

Note: For information on the difference of two squares method, see page 194.

Factor Simple Trinomials

Factoring simple trinomials involves breaking down an expression with three terms into its smaller parts, known as factors.

If you have a quadratic trinomial that has a coefficient of one — in other words, a trinomial whose highest exponent is 2 and has no number multiplying the first variable—you may be able to use the trinomial method of factoring.

Quadratic trinomials look like $x^2 + bx + c$ when the terms are written in order of decreasing

powers and $(x + \_)(x + \_)$ when they are factored out. To fill in the blanks, you must do a little detective work to find two numbers that equal b when they are added together and equal c when multiplied together.

Taking note of whether b and c are positive or negative will help you determine whether the two answers are both positive, both negative, or one of each. The two numbers you arrive at will fill in the blanks and give you the factors for your original expression.

Factor Simple Trinomials Example 1

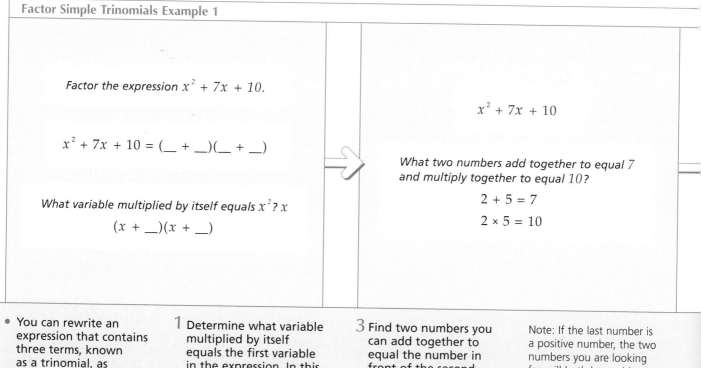

Factor the expression $x^2 + 7x + 10$.

$x^2 + 7x + 10 = (\_ + \_)(\_ + \_)$

What variable multiplied by itself equals x^2? x

$(x + \_)(x + \_)$

$x^2 + 7x + 10$

What two numbers add together to equal 7 and multiply together to equal 10?

$2 + 5 = 7$

$2 \times 5 = 10$

- You can rewrite an expression that contains three terms, known as a trinomial, as $(\_ + \_)(\_ + \_)$. You need to determine what numbers and/or variables go into each blank spot.

- You can only use this method if a number does not appear in front of the first variable.

1 Determine what variable multiplied by itself equals the first variable in the expression. In this example, $x \times x$ equals x^2.

2 Place the variable in the first position of both sets of parentheses. In this example, write $(x + \_)(x + \_)$.

3 Find two numbers you can add together to equal the number in front of the second variable and multiply together to equal the last number. In this example, $2 + 5$ equals 7 and 2×5 equals 10.

Note: If the last number is a positive number, the two numbers you are looking for will both be positive or negative numbers. If the last number is a negative number, one number will be a positive number and one number will be a negative number.

Tip

When an expression is in the form $x^2 + bx + c$, how can I determine which numbers add to equal b and multiply to equal c?

You can start by writing down all the numbers that multiply together to equal c in the expression. You can then see which of these numbers add up to give you the value of b. For example, in the expression $x^2 + 7x + 10$, you need to find the numbers that multiply together to equal 10 and add together to equal 7.

$$1 \times 10 = 10 \qquad 1 + 10 = 11$$
$$2 \times 5 = 10 \qquad 2 + 5 = 7$$

Practice

$a+b=c$

Factor the following expressions. You can check your answers on page 263.

1) $x^2 + 6x + 5$

2) $x^2 + 12x + 20$

3) $x^2 - x - 12$

4) $x^2 - 5x + 6$

5) $x^2 - 6x + 8$

6) $x^2 + 8x - 9$

$$x^2 + 7x + 10 = (x + 2)(x + 5)$$

Check Your Answer
$(x + 2)(x + 5)$
$= x^2 + 5x + 2x + 10$
$- x^2 + 7x + 10$ *Correct!*

Factor Simple Trinomials Example 2

Factor the expression $x^2 - 3x - 18$.
$$x^2 - 3x - 18 = (\_ + \_)(\_ + \_)$$

What variable multiplied by itself equals x^2? x
$$(x + \_)(x + \_)$$

What two numbers add together to equal -3 and multiply together to equal -18?
$$3 + (-6) = -3$$
$$3 \times (-6) = -18$$

$$x^2 - 3x - 18 = (x + 3)(x - 6)$$

---

4 Place the numbers you determined in step 3 into the remaining positions in the parentheses ( ), in any order.

Note: In this example, you could also write the result as $(x + 5)(x + 2)$.

5 To check your answer, multiply each term in the first set of parentheses by each term in the second set of parentheses.

• If you end up with the original expression, you have correctly factored the expression.

Note: For more information on multiplying terms in parentheses, see page 154.

1 Determine what variable multiplied by itself equals the first variable in the expression.

2 Place the variable in the first position of both sets of parentheses.

3 Find two numbers you can add together to equal the number in front of the second variable and multiply together to equal the last number. In this example, $3 + (-6)$ equals -3 and $3 \times (-6)$ equals -18.

4 Place the numbers into the remaining positions in the parentheses ( ), in any order.

• You have finished factoring the expression.

Factor More Complex Trinomials

Factoring a trinomial involves breaking a three-term expression down into smaller parts, known as factors. The method of factoring shown below allows you to factor a quadratic trinomial—in other words, a trinomial whose highest exponent is 2. If a number does not appear in front of the first variable, you can see page 198 for an easier way to factor the expression.

Quadratic trinomials look like $ax^2 + bx + c$ when the terms are written in order of decreasing powers and $(x + \_\_)(x + \_\_)$ when they are factored out. To fill in the blanks, you must find two magic numbers that equal b when they are added together and equal

the product of ac when multiplied together. Start by writing down all the numbers that multiply together to equal the product of ac and then select the numbers that add up to the value of b. If no such combination of numbers exists, the expression cannot be factored and is known as a prime trinomial.

After determining the magic numbers you require, rewrite the equation to replace b with the two numbers added together. Then you can simply factor the expression using the grouping method discussed on page 192.

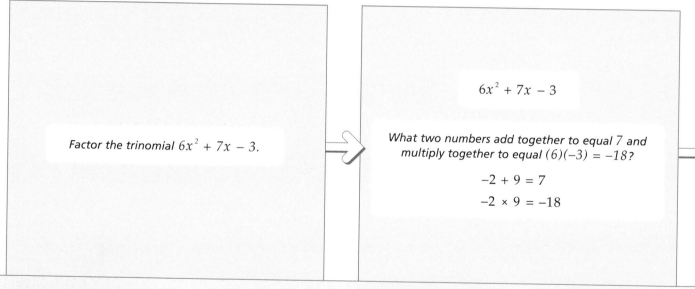

Factor the trinomial $6x^2 + 7x - 3$.

$6x^2 + 7x - 3$

What two numbers add together to equal 7 and multiply together to equal $(6)(-3) = -18$?

$$-2 + 9 = 7$$
$$-2 \times 9 = -18$$

- You can use this method to factor an expression with three terms, known as a trinomial, when the three terms are in the form $ax^2 + bx + c$, where a, b and c are numbers.

Note: If a number does not appear in front of the first variable, see page 198 for an easier way to factor the expression.

1 Find two numbers you can add together to equal the number in front of the second variable. These two numbers must also multiply together to equal the result of multiplying the number in front of the first variable by the last number.

- In this example, you need to find two numbers that add together to equal 7 and multiply together to equal $(6)(-3) = -18$. In this case, -2 and 9 are the numbers you need.

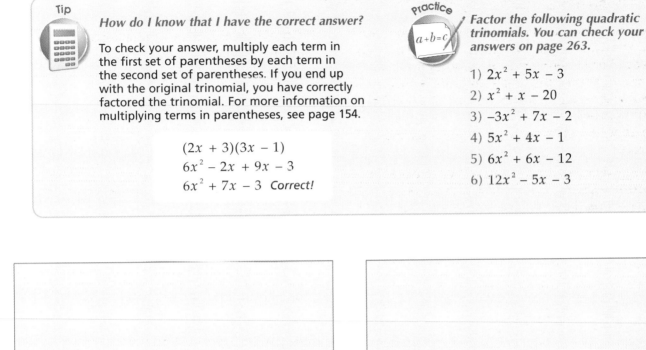

How do I know that I have the correct answer?

To check your answer, multiply each term in the first set of parentheses by each term in the second set of parentheses. If you end up with the original trinomial, you have correctly factored the trinomial. For more information on multiplying terms in parentheses, see page 154.

$$(2x + 3)(3x - 1)$$
$$6x^2 - 2x + 9x - 3$$
$$6x^2 + 7x - 3 \quad Correct!$$

Practice

Factor the following quadratic trinomials. You can check your answers on page 263.

1) $2x^2 + 5x - 3$

2) $x^2 + x - 20$

3) $-3x^2 + 7x - 2$

4) $5x^2 + 4x - 1$

5) $6x^2 + 6x - 12$

6) $12x^2 - 5x - 3$

$$6x^2 + 7x - 3$$
$$6x^2 + (-2 + 9)x - 3$$
$$6x^2 - 2x + 9x - 3$$

Group 1   Group 2

$$6x^2 - 2x + 9x - 3$$
$$2x(3x - 1) + 3(3x - 1)$$
$$(2x + 3)(3x - 1)$$

2 Replace the number in front of the second variable with the numbers you determined in step 1. Write the numbers as the sum of the two numbers, in any order, and surround the numbers with parentheses ( ).

• In this example, replace the 7 in $7x$ with $(-2 + 9)$.

3 To remove the parentheses, multiply each number in the parentheses by the x variable.

4 To finish factoring the trinomial, use the factor by grouping technique.

Note: To use the factor by grouping technique to factor an expression, see page 192.

• When you finish factoring the trinomial, you will end up with two sets of terms, with two terms each, with each set of terms surrounded by parentheses.

Factoring Expressions

Question 1. Factor the following expressions using the greatest common factor method.

 a) $3 + 6 + 15$

 b) $12 + 8 + 20$

 c) $8x^2 + 2x + 6x^3$

 d) $a^2bc + ab^2c + abc^2$

 e) $6a^3b^2 + 9a^2b + 21a^4b^3$

Question 2. Factor the following expressions using the grouping method.

 a) $x^2 + 3x + 4x + 12$

 b) $4x^2 - 2x + 6x - 3$

 c) $3y^2 - 12y + 5y - 20$

 d) $x^2 + 2x + x + 2$

 e) $xy - 3y + x - 3$

 f) $2xy + 2y^2 - xy - x^2$

Question 3. Factor the following quadratic expressions.

a) $x^2 + 5x + 4$

b) $x^2 + x - 6$

c) $z^2 - 3z - 10$

d) $y^2 - 9$

e) $a^2 + 9a + 14$

Question 4. Factor the following quadratic expressions.

a) $2x^2 + 3x - 2$

b) $6x^2 + 7x + 2$

c) $3x^2 - 7x - 20$

d) $4x^2 - 12x + 5$

e) $20x^2 + 11x - 3$

Question 5. Factor the following expressions as much as possible.

a) $x^2 - y^2$

b) $a^3 - b^3$

c) $x^3 + 8$

d) $y^3 - 1$

e) $x^4 - 1$

You can check your answers on page 278.

$$x^2 - 9 = 0$$

Chapter 11

Solving quadratic equations and inequalities is a bit more tricky than solving basic equations and inequalities. Chapter 11 helps you recognize quadratic equations in the various forms in which they can appear and helps you decide on the best way to tackle them. This chapter will also help you solve quadratic inequalities.

$$x^2 -$$

$$5x^2$$

Solving Quadratic Equations and Inequalities

In this Chapter...

Solve Quadratic Equations by Factoring

Method 1

Once you have honed your factoring skills, they can be put to good use solving quadratic equations. A quadratic equation is an equation in which the highest exponent is 2 and is generally simplified down to the form $ax^2 + bx + c = 0$. Sometimes, however, you will run into an equation in which c is equal to 0. These equations have the form of $ax^2 + bx = 0$ and are simple to solve by factoring. For more information on factoring, see chapter 10.

One method of solving by factoring requires you to start by factoring the variable out of the equation

so the equation takes the form of $x(ax + b) = 0$. You should then separate each factor into an equation that equals zero, because if two factors multiply to equal zero, then at least one of the factors must equal zero. You will get one possible solution for x right off the bat: $x = 0$. Solve for the variable in the second equation to get the second possible solution for x.

You can check the possible solutions you found by plugging your answers into the original equation to see if they both satisfy the equation.

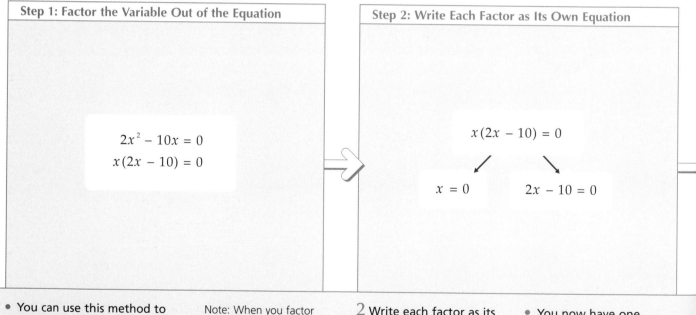

Step 1: Factor the Variable Out of the Equation

$$2x^2 - 10x = 0$$
$$x(2x - 10) = 0$$

Step 2: Write Each Factor as Its Own Equation

$$x(2x - 10) = 0$$

$$x = 0 \qquad 2x - 10 = 0$$

- You can use this method to solve a quadratic equation when the equation has only two terms in the form $ax^2 + bx = 0$, where a and b are numbers.

1 Factor the left side of the equation by factoring out the variable, such as x.

Note: When you factor an equation, you break the equation into pieces, called factors, which multiply together to produce the original equation. To factor a variable out of an equation, see page 188.

2 Write each factor as its own equation and set each factor equal to 0.

- You now have one solution to the problem. In this example, x can equal 0.

Note: $x = 0$ will always be one solution to the problem.

Tip

Is there a faster way to find the possible solutions for a quadratic equation?

If an equation is in the form $ax^2 + bx = 0$, where a and b are numbers, the two possible solutions will always be $x = 0$ and $x = -\frac{b}{a}$. For example, for the equation $2x^2 - 10x = 0$ below, the two solutions are $x = 0$ and $x = -(\frac{-10}{2}) = 5$—the same result that we obtained below. Using this trick to find the solutions can help you quickly solve quadratic equations.

Practice

Solve the following equations by factoring. You can check your answers on page 263.

1) $x^2 - 7x = 0$

2) $x^2 + 5x = 0$

3) $3x^2 - 8x = 0$

4) $2x^2 - 20x = 0$

5) $3x^2 + 3x = 2x - x^2$

6) $2x^2 - 4x + 5 = x^2 + 5$

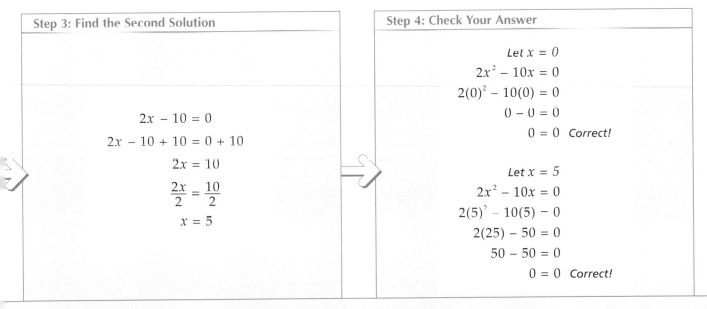

Step 3: Find the Second Solution

$$2x - 10 = 0$$
$$2x - 10 + 10 = 0 + 10$$
$$2x = 10$$
$$\frac{2x}{2} = \frac{10}{2}$$
$$x = 5$$

Step 4: Check Your Answer

Let $x = 0$
$$2x^2 - 10x = 0$$
$$2(0)^2 - 10(0) = 0$$
$$0 - 0 = 0$$
$$0 = 0 \quad \textit{Correct!}$$

Let $x = 5$
$$2x^2 - 10x = 0$$
$$2(5)^2 - 10(5) = 0$$
$$2(25) - 50 = 0$$
$$50 - 50 = 0$$
$$0 = 0 \quad \textit{Correct!}$$

3 To determine the second solution to the problem, determine which numbers you need to add, subtract, multiply and/or divide to place the variable by itself on the left side of the equation.

4 Add, subtract, multiply and/or divide both sides of the equation by the numbers you determined in step 3.

• In this example, add 10 to both sides of the equation. Then divide both sides of the equation by 2.

Note: For more information on adding, subtracting, multiplying and dividing numbers in equations, see pages 64 to 67.

• You have found the second solution to the original equation. In this example, x can equal 5.

5 To check your answer, place the numbers you found into the original equation and solve the problems. If both sides of the equation are equal in both cases, you have correctly solved the equation.

• In this example, the solutions to the equation are 0 and 5.

CONTINUED

Solve Quadratic Equations by Factoring *continued*

Method 2

As you have seen, quadratic equations are generally simplified down to the form $ax^2 + bx + c = 0$. While there are different ways of solving different types of quadratic equations, factoring is a foolproof method you can use to solve all types.

To begin using the method of solving by factoring shown below, set one side of the quadratic equation to equal zero by moving all the terms to the left side of the equation. Then you can factor the equation. Once you have the equation down to two factors, you can split the equation into two

equations with one side of each equation set to equal zero. With the two smaller equations you now have, solving for x is straightforward.

In some cases, you may end up with two factors that are identical, such as $(x + 2)(x + 2) = 0$. This leads to two identical solutions, which is called a double root.

Again, it is important to check that you have not made an error in your calculations. Plug your answers into the original equation to see if they are correct.

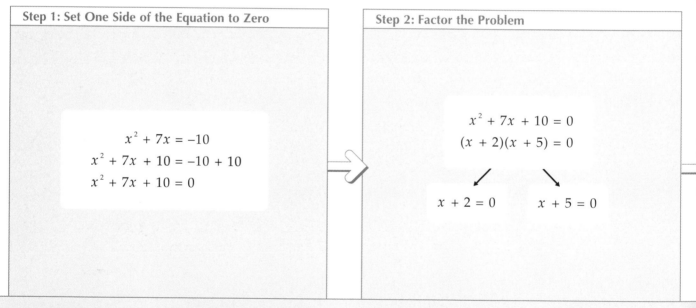

Step 1: Set One Side of the Equation to Zero

$$x^2 + 7x = -10$$
$$x^2 + 7x + 10 = -10 + 10$$
$$x^2 + 7x + 10 = 0$$

Step 2: Factor the Problem

$$x^2 + 7x + 10 = 0$$
$$(x + 2)(x + 5) = 0$$

$$x + 2 = 0 \qquad x + 5 = 0$$

- You can use this method to solve any quadratic equation.

1 Determine which numbers and variables you need to add and/or subtract to place all the numbers and variables on the left side of the equation and make the right side of the equation equal to 0.

2 Add and/or subtract the numbers and variables you determined in step **1** on both sides of the equation.

- In this example, add 10 to both sides of the equation.

Note: For more information on adding and subtracting numbers and variables in equations, see page 64.

3 Factor the left side of the equation.

Note: When you factor a problem, you break the problem into pieces, called factors, which multiply together to produce the original problem. You can use the techniques you learned in chapter 10 to factor a problem. In this example, see page 198 to factor the problem.

4 Write each factor as its own equation and set each factor equal to 0.

Tip

After factoring an equation, I ended up with a factor that is a number. What should I do?

You can ignore any factor that is a number because these factors are so easy to eliminate. Just divide both sides of the equation by the number and it disappears. For example, when you factor the equation $4x^2 + 28x + 40 = 0$, you end up with the equation $4(x + 2)(x + 5) = 0$. Dividing both sides by 4 leaves you with $(x + 2)(x + 5) = 0$.

Practice

$a+b=c$

Solve for the variable by factoring in the following equations. You can check your answers on page 263.

1) $x^2 - x - 2 = 0$

2) $2x^2 - 5x - 3 = 0$

3) $6x^2 - 19x + 10 = 0$

4) $x^2 + 8x - 10 = 7x + 10$

5) $2x^2 + 5x = -2x^2 + 8x + 1$

6) $x^2 - 2x + 2 - 2x - 2$

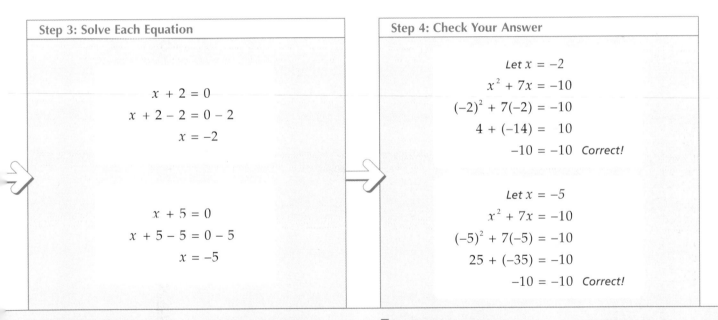

Step 3: Solve Each Equation

$$x + 2 = 0$$
$$x + 2 - 2 = 0 - 2$$
$$x = -2$$

$$x + 5 = 0$$
$$x + 5 - 5 = 0 - 5$$
$$x = -5$$

Step 4: Check Your Answer

Let $x = -2$
$$x^2 + 7x = -10$$
$$(-2)^2 + 7(-2) = -10$$
$$4 + (-14) = 10$$
$$-10 = -10 \quad Correct!$$

Let $x = -5$
$$x^2 + 7x = -10$$
$$(-5)^2 + 7(-5) = -10$$
$$25 + (-35) = -10$$
$$-10 = -10 \quad Correct!$$

5 In each equation, determine which numbers you need to add, subtract, multiply and/or divide to place the variable by itself on the left side of the equation.

6 Add, subtract, multiply and/or divide both sides of the equation by the numbers you determined in step 5.

- In this example, subtract 2 from both sides of the first equation. Subtract 5 from both sides of the second equation.

- Each answer is a solution to the original equation.

7 To check your answer, place the numbers you found into the original equation and solve the problems. If both sides of the equation are equal in both cases, you have correctly solved the equation.

- In this example, the solutions to the equation are −2 and −5.

Solve Quadratic Equations Using Square Roots

While quadratic equations are generally simplified down to the form of $ax^2 + bx + c = 0$, you will sometimes run into an equation in which b is equal to 0. These equations only have two terms and have the form of $ax^2 + c = 0$. You can use square roots to solve for the variable in quadratic equations in this form.

Using square roots allows you to easily solve for a variable. All you have to do is add, subtract, multiply and divide as necessary to isolate the squared variable, usually x^2, to one side of the equation. Then, you must remove the exponent by finding the square root of both sides of the equation. When solving quadratic equations, however, the negative value of the square root must also be considered as a solution. For example, if your equation is $x^2 = 16$, you will arrive at a solution of $x = \pm 4$. The plus/minus symbol ($\pm$) represents two possible solutions for x: 4 and –4.

To check if your solutions are correct, plug each number into the original equation and work out the equation.

Step 1: Isolate the Variable

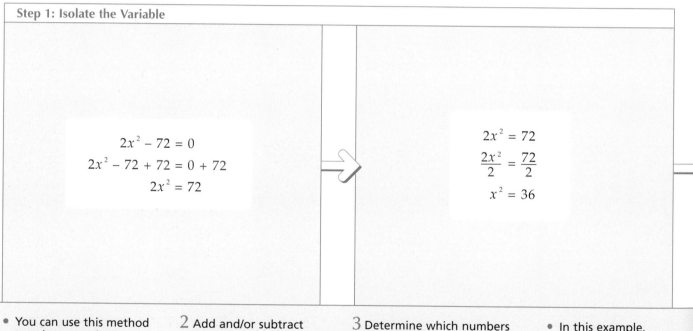

$$2x^2 - 72 = 0$$
$$2x^2 - 72 + 72 = 0 + 72$$
$$2x^2 = 72$$

$$2x^2 = 72$$
$$\frac{2x^2}{2} = \frac{72}{2}$$
$$x^2 = 36$$

- You can use this method to solve a quadratic equation when the equation has only two terms in the form $ax^2 + c = 0$, where a and c are numbers.

1 Determine which numbers you need to add and/or subtract to place the variable by itself on the left side of the equation.

2 Add and/or subtract the numbers you determined in step **1** on both sides of the equation.

- In this example, add 72 to both sides of the equation.

Note: For more information on adding and subtracting numbers in equations, see page 64.

3 Determine which numbers you need to multiply and/or divide to place the variable by itself on the left side of the equation.

4 Multiply and/or divide both sides of the equation by the numbers you determined in step **3**.

- In this example, divide both sides of the equation by 2.

Note: For more information on multiplying and dividing numbers in equations, see page 66.

Tip

Can I only use this method to solve equations written in the form $ax^2 + c = 0$?

Keep in mind that the same equation can be written in many different forms. As long as you can put the equation into the form $ax^2 + c = 0$, you can use this method to solve the equation. For example, you could use this method to solve the equations $2x^2 = 72$ and $2x^2 - 10 = 62$, since these equations are the same as $2x^2 - 72 = 0$.

Practice

Solve for the variable in the following equations. You can check your answers on page 263.

1) $x^2 - 9 = 0$

2) $x^2 - 25 = 0$

3) $4x^2 - 400 = 0$

4) $3x^2 + 10 = 58$

5) $x^2 + 6 = 14 - x^2$

6) $36 = x^2$

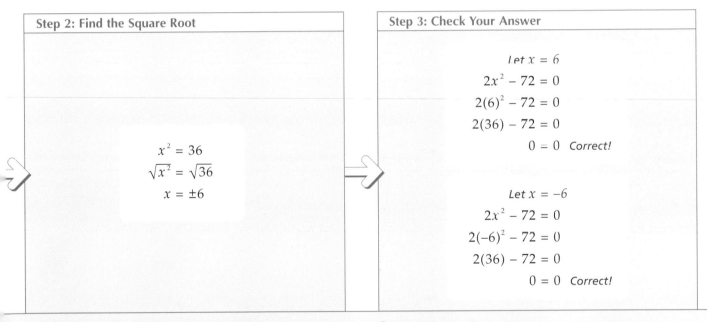

Step 2: Find the Square Root	Step 3: Check Your Answer
$x^2 = 36$ $\sqrt{x^2} = \sqrt{36}$ $x = \pm 6$	$Let\ x = 6$ $2x^2 - 72 = 0$ $2(6)^2 - 72 = 0$ $2(36) - 72 = 0$ $0 = 0$ *Correct!* $Let\ x = -6$ $2x^2 - 72 = 0$ $2(-6)^2 - 72 = 0$ $2(36) - 72 = 0$ $0 = 0$ *Correct!*

5 To remove the exponent, find the square root of both sides of the equation.

Note: A radical sign ($\sqrt{}$) indicates that you want to find the root of a variable or number. To find the square root of a variable or number, ask yourself "What variable or number multiplied by itself equals the variable or number under the radical sign?" For more information on radicals, see page 168.

• When you find the square root of a number, you need to place the $\pm$ symbol before the number to indicate that the answer could be a positive or negative number.

6 To check your answer, place the numbers you found into the original equation and solve the problems. If both sides of the equation are equal in both cases, you have correctly solved the equation.

• In this example, the solutions to the equation are 6 and −6.

Solve Quadratic Equations by Completing the Square

One sure-fire way to find a solution for a quadratic equation—an equation in which the highest exponent is 2 and usually in the form of $ax^2 + bx + c = 0$—is the completing the square method. The completing the square method takes a quadratic equation and forces it to become an expression that contains a square, making the equation much easier to solve.

To convert a quadratic equation into an equation that contains a square, you have to prepare the equation. First, eliminate the number, or coefficient, in front of x^2. Then make some room at the end of the equation by moving the number that is not attached to a variable to the other side of the equals sign. Next comes the tricky part. Take the number in front of x, divide the number in half and then square the resulting number. Adding the resulting number to the end of the equation will give you an equation that is a square and can be rewritten in the form of $(x + a)^2$.

Step 1: Simplify the Equation

Solve the equation $2x^2 + 12x - 14 = 0$.

$$2x^2 + 12x - 14 = 0$$
$$\frac{2x^2}{2} + \frac{12x}{2} - \frac{14}{2} = \frac{0}{2}$$
$$x^2 + 6x - 7 = 0$$

Step 2: Isolate the Variables

$$x^2 + 6x - 7 = 0$$
$$x^2 + 6x - 7 + 7 = 0 + 7$$
$$x^2 + 6x = 7$$

- To solve a quadratic equation by completing the square, you will first want to remove the number in front of the x^2 term.

1 If a number appears in front of the x^2 term, divide each term in the equation by the number in front of the x^2 term.

- In this example, the number 2 appears in front of the x^2 term, so divide each term in the equation by 2.

2 Determine which numbers and/or variables you need to add or subtract to place the variables on the left side of the equation and the numbers on the right side of the equation.

3 Add and subtract the numbers and/or variables you determined in step 2 on both sides of the equation.

- In this example, add 7 to both sides of the equation.

Note: For more information on adding and subtracting numbers and variables in equations, see page 64.

Tip

How can I make sure that I correctly wrote the left side of the equation in squared form?

You can do a quick check by multiplying out the rewritten form of the equation. If you end up with the same equation you had before rewriting, you have the correct squared form of the equation. For example, if you determine the squared form to be $(x + 3)^2$, multiply $(x + 3)$ by $(x + 3)$ to see if you end up with the original terms, which in this case are $x^2 + 6x + 9$. To multiply polynomials, you multiply each term in the first set of parentheses by each term in the second set of parentheses.

Step 3: Write the Left Side of the Equation in Squared Form

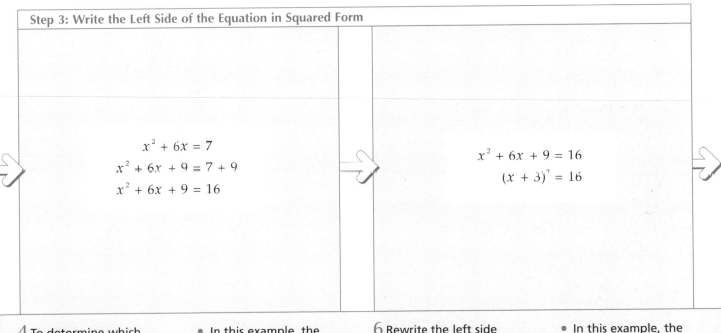

$$x^2 + 6x = 7$$
$$x^2 + 6x + 9 = 7 + 9$$
$$x^2 + 6x + 9 = 16$$

$$x^2 + 6x + 9 = 16$$
$$(x + 3)^2 = 16$$

4 To determine which number you need to add to both sides of the equation so you can write the left side of the equation in squared form, divide the number in front of the x term by 2 and then multiply the resulting number by itself.

- In this example, the number in front of the x term is 6, so divide 6 by 2, which equals 3. Then multiply 3 by 3, which equals 9.

5 Add the number you determined in step 4 to both sides of the equation.

6 Rewrite the left side of the equation in the form $(x + a)^2$, where a is equal to the number in front of the x term, divided by 2.

- In this example, the number in front of the x term is 6, so divide 6 by 2, which equals 3. In this case, a equals 3.

CONTINUED

Solve Quadratic Equations by Completing the Square *continued*

When solving a quadratic equation using the completing the square method, once you have an expression that is a perfect square, you must remove the exponent by finding the square root of both sides of the equation. For example, if your equation is $(x + 2)^2 = 16$, find the square root of both sides, giving you $x + 2 = \pm\sqrt{16}$, or ± 4. The plus/minus symbol ($\pm$) represents two answers—the positive and negative value of the number—and is very important because square roots typically have two answers. For example, 4 and –4 are both square roots of 16.

Next, you simply isolate the variable to give you the two solutions for the quadratic equation. You may end up with a radical in your answer. For example, you may have $x = -\sqrt{5} - 2$ and $x = \sqrt{5} - 2$. That's just fine and you can leave your answer in that form. Most teachers prefer to see a nice neat radical as opposed to a number with messy decimal places.

With some practice, the completing the square technique will become automatic for you. Remember to check your answers by placing them into the original equation and working out the problem.

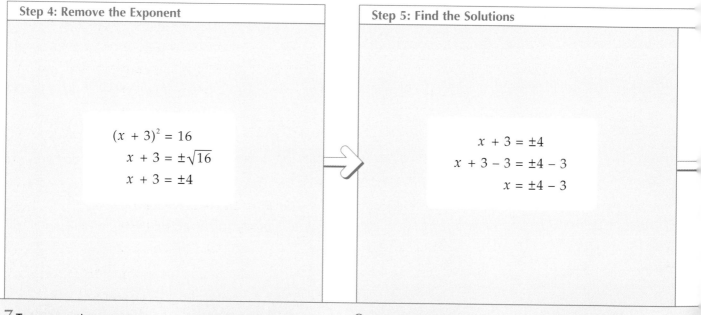

Step 4: Remove the Exponent

$$(x + 3)^2 = 16$$
$$x + 3 = \pm\sqrt{16}$$
$$x + 3 = \pm 4$$

Step 5: Find the Solutions

$$x + 3 = \pm 4$$
$$x + 3 - 3 = \pm 4 - 3$$
$$x = \pm 4 - 3$$

7 To remove the exponent from the left side of the equation, find the square root of both sides of the equation.

Note: A radical sign ($\sqrt{}$) indicates that you want to find the root of a number. For more information on radicals, see page 168.

- When you find the square root of a number, you need to place the $\pm$ symbol before the number to indicate that the answer could be a positive or negative number.

8 Determine which number you need to add or subtract to place the variable by itself on the left side of the equation.

9 Add or subtract the number you determined in step 8 on both sides of the equation.

- In this example, subtract 3 from both sides of the equation.

Practice

$a + b = c$

Solve the following quadratic equations by completing the square. You can check your answers on page 263.

1) $4x^2 + 8x - 12 = 0$

2) $x^2 - 2x + 1 = 0$

3) $x^2 - 4x + 7 = 0$

4) $2x^2 + 8x - 10 = 0$

5) $x^2 - 1 = 0$

6) $x^2 + x - 6 = 0$

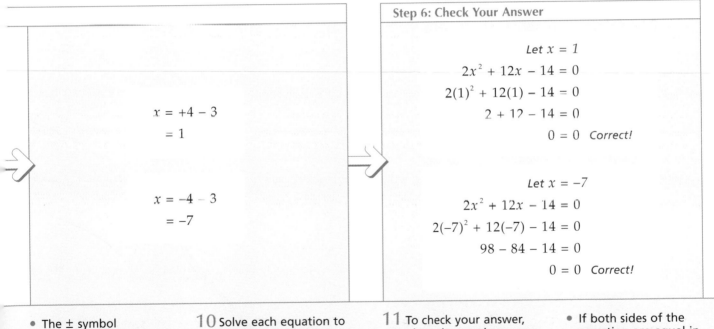

$x = +4 - 3$

$= 1$

$x = -4 - 3$

$= -7$

Step 6: Check Your Answer

$$\text{Let } x = 1$$

$$2x^2 + 12x - 14 = 0$$

$$2(1)^2 + 12(1) - 14 = 0$$

$$2 + 12 - 14 = 0$$

$$0 = 0 \quad \textit{Correct!}$$

$$\text{Let } x = -7$$

$$2x^2 + 12x - 14 = 0$$

$$2(-7)^2 + 12(-7) - 14 = 0$$

$$98 - 84 - 14 = 0$$

$$0 = 0 \quad \textit{Correct!}$$

- The ± symbol indicates that you have two solutions, one with a plus sign (+) and the other with a minus sign (−).

10 Solve each equation to find the two solutions to the equation.

- In this example, the solutions to the equation are 1 and −7.

11 To check your answer, place the numbers you found into the original equation and solve the problems.

- If both sides of the equation are equal in both cases, you have correctly solved the equation.

Solve Quadratic Equations Using the Quadratic Formula

A quadratic equation is an equation that can be simplified down to the form of $ax^2 + bx + c = 0$. The highest exponent occurring in a quadratic equation is 2. You can use the quadratic formula to help you solve quadratic equations. The quadratic formula is well worth committing to memory as it is easy to use and can be applied to any quadratic equation.

The quadratic formula does, however, have a couple of quirks. First, the formula contains a radical. Working out the value under the radical sign is your first task. For information on radicals, see page 168. Second, the formula contains the plus/minus symbol (±). This symbol eventually splits the solution into two equations which will provide you with the two possible answers for a quadratic equation.

As always, check your two solutions for x by plugging them into the original equation.

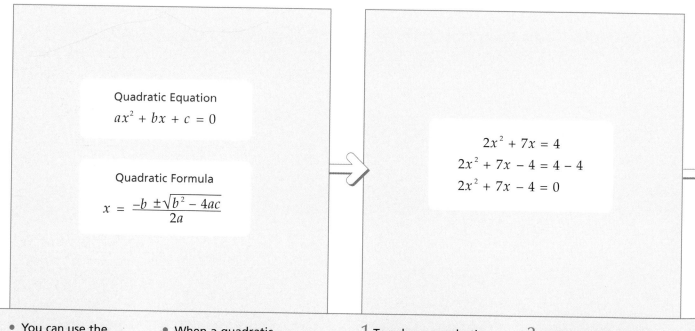

Quadratic Equation
$$ax^2 + bx + c = 0$$

Quadratic Formula
$$x = \frac{-b \pm \sqrt{b^2 - 4ac}}{2a}$$

$$2x^2 + 7x = 4$$
$$2x^2 + 7x - 4 = 4 - 4$$
$$2x^2 + 7x - 4 = 0$$

- You can use the quadratic formula to solve any quadratic equation.

  Note: A quadratic equation is an equation whose highest exponent is two, such as x^2.

- When a quadratic equation is written in the form $ax^2 + bx + c = 0$, you use the numbers in the equation, represented by a, b and c, in the quadratic formula.

- For example, in the equation $2x^2 + 7x - 4 = 0$, a equals 2, b equals 7 and c equals –4.

1 To solve a quadratic equation, determine if the equation is written in the standard form of $ax^2 + bx + c = 0$.

2 If the equation is not written in the standard form, determine which numbers you need to add or subtract to write the equation in standard form.

3 Add or subtract the numbers you determined in step 2 on both sides of the equation.

- In this example, subtract 4 from both sides of the equation.

  Note: For more information on adding and subtracting numbers in equations, see page 64.

Tip

What can I do if bx or c is missing from the quadratic equation?

Quadratic equations are usually written in the form $ax^2 + bx + c = 0$, but the second or third term may be missing if b or c equals 0. For example, in the equation $2x^2 + 10 = 0$, b equals 0, while in the equation $2x^2 + 4x = 0$, c equals 0. You can still use the quadratic formula to solve these equations. Simply plug any 0s into the formula along with the other values. If ax^2 is missing, however, the equation is not quadratic and the formula does not apply.

Practice

$a + b = c$

Solve the following equations using the quadratic formula. You can check your answers on page 264.

1) $x^2 + x - 2 = 0$

2) $x^2 - 9 = 0$

3) $x^2 - 2x + 2 = 0$

4) $12x^2 + 5x - 2 = 0$

5) $5x^2 - 7x = 0$

6) $x^2 - 4x + 4 = 0$

$2x^2 + 7x - 4 = 0$

$a = 2, b = 7, c = -4$

$x = \dfrac{-b \pm \sqrt{b^2 - 4ac}}{2a}$

$x = \dfrac{-7 \pm \sqrt{7^2 - 4(2)(-4)}}{2(2)}$

$x = \dfrac{-7 \pm \sqrt{49 - (-32)}}{4}$

$x = \dfrac{-7 \pm \sqrt{81}}{4}$

$x = \dfrac{-7 + \sqrt{81}}{4}$ $x = \dfrac{-7 - \sqrt{81}}{4}$

$x = \dfrac{-7 + 9}{4}$ $x = \dfrac{-7 - 9}{4}$

$x = \dfrac{2}{4}$ $x = \dfrac{-16}{4}$

$x = \dfrac{1}{2}$ $x = -4$

---

4 Determine the values of a, b and c in the equation.

Note: If a number does not appear before x^2 or x, assume the number is 1.

5 Place the numbers you determined for a, b and c into the quadratic formula.

6 Solve the problem under the radical sign ($\sqrt{}$).

Note: Make sure you calculate b^2 first, then multiply 4, a and c together and then subtract the results. When you see two negative signs (–) next to each other, you can change the negative signs to one positive sign (+). For example, $49 - (-32)$ equals $49 + 32$.

• The ± symbol in the quadratic formula indicates that the equation can be broken into two separate equations, one with a plus sign (+) and the other with a minus sign (–).

7 Separate the equation into two equations, one with a plus sign (+) and one with a minus sign (–) before the radical sign ($\sqrt{}$). Solve each equation separately.

Note: Each equation contains a radical sign ($\sqrt{}$), which tells you to find the square root of a number. For more information on radicals, see page 168.

• In this example, the possible solutions to the equation are $\frac{1}{2}$ and -4.

Solve Cubic and Higher Equations

Finding solutions for large equations with variables raised to the 3rd power or higher might be a little more complex than solving quadratic equations, but you can do it, mainly using techniques you are already familiar with.

To find possible solutions for the equation, look at the number at the end of the equation that is not attached to a variable and at the number multiplying the variable with the highest exponent and create a list of factors for both of these numbers. Then divide each of the factors for the number at the end of the equation by each of the factors for the number multiplying the variable with the highest exponent. After eliminating duplicates and simplifying all the fractions, you will end up with a handy list of possible solutions for the equation.

To separate the real solutions from the impostors, you can plug one of the possible solutions into the equation. If both sides of the equation are equal, you have found one solution to the equation. You can then use synthetic division to help you find the other solutions. For more information on synthetic division, see page 160.

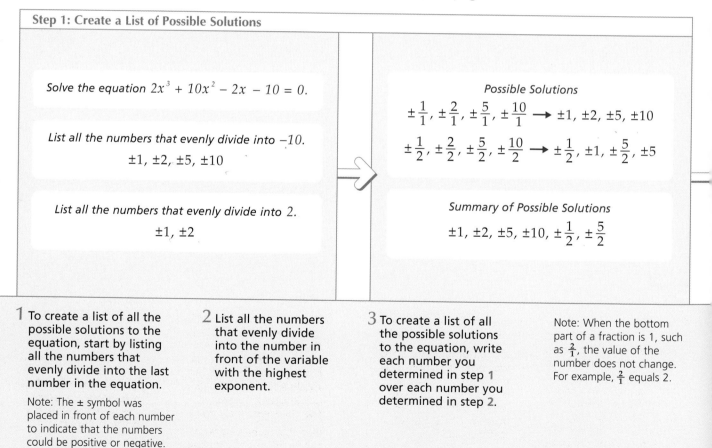

Step 1: Create a List of Possible Solutions

Solve the equation $2x^3 + 10x^2 - 2x - 10 = 0$.

List all the numbers that evenly divide into -10.
$\pm 1, \pm 2, \pm 5, \pm 10$

List all the numbers that evenly divide into 2.
$\pm 1, \pm 2$

Possible Solutions

$\pm \frac{1}{1}, \pm \frac{2}{1}, \pm \frac{5}{1}, \pm \frac{10}{1} \longrightarrow \pm 1, \pm 2, \pm 5, \pm 10$

$\pm \frac{1}{2}, \pm \frac{2}{2}, \pm \frac{5}{2}, \pm \frac{10}{2} \longrightarrow \pm \frac{1}{2}, \pm 1, \pm \frac{5}{2}, \pm 5$

Summary of Possible Solutions

$\pm 1, \pm 2, \pm 5, \pm 10, \pm \frac{1}{2}, \pm \frac{5}{2}$

1 To create a list of all the possible solutions to the equation, start by listing all the numbers that evenly divide into the last number in the equation.

Note: The ± symbol was placed in front of each number to indicate that the numbers could be positive or negative. For example, ±1 indicates that the number could be +1 or −1.

2 List all the numbers that evenly divide into the number in front of the variable with the highest exponent.

3 To create a list of all the possible solutions to the equation, write each number you determined in step **1** over each number you determined in step **2**.

Note: When the bottom part of a fraction is 1, such as $\frac{2}{1}$, the value of the number does not change. For example, $\frac{2}{1}$ equals 2.

Tip

How many solutions will an equation have?

An equation will have up to the same number of solutions as the highest exponent in the equation. An equation may have fewer solutions, but it will never have more. For example, an equation containing x^3 as the variable with the highest exponent, may have three, two or one solutions.

Tip

Is there a quicker way to solve cubic or higher equations?

If an equation has a variable raised to an exponent of 3 or higher, you may be able to solve the equation using the same factoring method you learned to solve quadratic equations (see page 206). For example, in the equation $x^3 + x^2 - 6x = 0$, you could factor out the x variable to end up with $x(x^2 + x - 6) = 0$. This would give you x = 0 as one solution to the equation. The remaining equation, $(x^2 + x - 6) = 0$, is a quadratic equation that you can simply factor to find the other possible solutions to the equation.

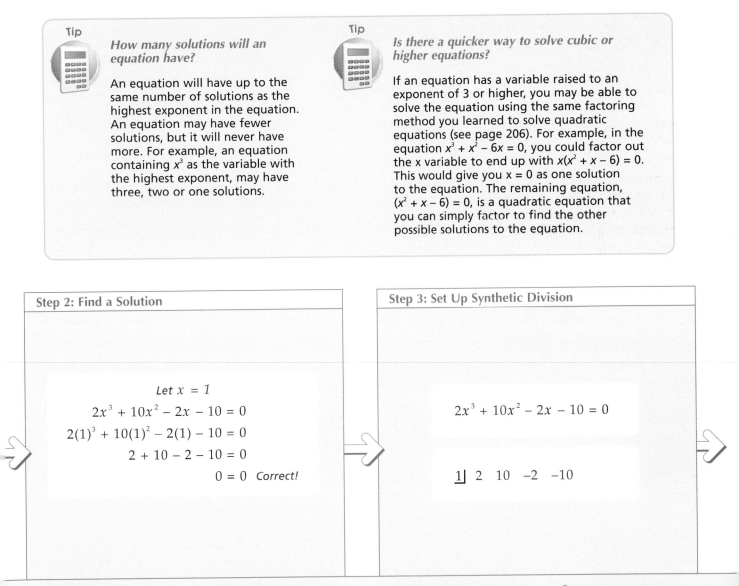

Step 2: Find a Solution

$$Let\ x = 1$$
$$2x^3 + 10x^2 - 2x - 10 = 0$$
$$2(1)^3 + 10(1)^2 - 2(1) - 10 = 0$$
$$2 + 10 - 2 - 10 = 0$$
$$0 = 0\quad Correct!$$

Step 3: Set Up Synthetic Division

$$2x^3 + 10x^2 - 2x - 10 = 0$$

$$\underline{1|}\ \ 2\ \ \ 10\ \ \ {-2}\ \ \ {-10}$$

4 To determine if one of the numbers you found in step 3 is a solution to the equation, place one of the numbers into the original equation and solve the problem. If both sides of the equation are equal, you have found one solution to the equation.

- In this example, 1 is a solution to the equation.

- If the number you tried was not a solution to the equation, try each number you found in step 3, one at a time, until you find a solution to the equation.

5 To find the other solutions to the equation, look at the numbers directly in front of the variables in the equation as well as the last number in the equation and write the numbers in a row.

Note: If a number does not appear in front of a variable, assume the number is 1. For example, x equals 1x.

6 Write the solution to the equation that you found in step 4 to the left of the row of numbers. Draw a partial box around the number to separate the number from the other numbers.

CONTINUED

After you have found one solution to the equation, you can perform synthetic division using the solution you found and the equation. For information on synthetic division, see page 160. Once you have a result for the synthetic division, you can rewrite the result as a new expression, which you then turn into an equation. If that equation is a quadratic equation, you can solve the equation using one of the factoring

techniques discussed in Chapter 10. Solving the quadratic equation will give you the remaining solutions for the original equation.

After you find all the solutions for the original equation, remember to check the numbers you determined to be solutions by plugging them into the original equation and working out the equation.

Step 4: Perform Synthetic Division

Step 5: Find Other Solutions

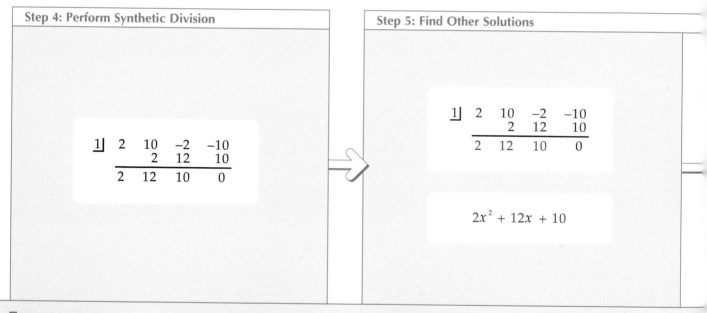

7 Using synthetic division, divide the number in the partial box into the row of numbers.

Note: For information on synthetic division, see page 160.

- The numbers below the line indicate the result of the division, with the last number being the remainder, or left over value.

- When a number is a solution to the equation, the remainder will be 0.

8 To find the other solutions to the equation, write the result of the division as an expression, writing each of the numbers below the line with decreasing powers of x next to the numbers. Start with the exponent one less than the highest exponent in the original equation.

- In this example, since the original equation starts with x^3, the new expression starts with x^2.

Tip

After I finish the synthetic division, what if I end up with an equation that is not a quadratic equation?

When you are solving an equation that contains a variable raised to an exponent of 4 or higher, such as x^4 or x^5, after you find one solution and perform synthetic division, you will end up with an equation that will not be a quadratic equation. You will have to repeat all the steps again with the new equation. Each time you follow these steps, you will reduce the highest exponent of the equation by one. Simply keep repeating this process until you end up with a quadratic equation.

Practice

Solve the following equations. You can check your answers on page 264.

1) $x^3 - 6x^2 + 11x - 6 = 0$

2) $3x^3 - 9x^2 - 27x = 15$

3) $x^3 - 2x^2 = 8 - 4x$

4) $2x^3 - 3x^2 = -1$

5) $x^4 - 2x^3 + x^2 = 0$

6) $x^4 + 5x^3 = x^2 + 5x$

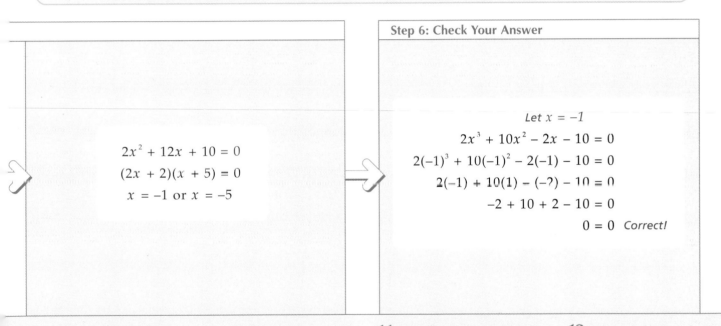

$$2x^2 + 12x + 10 = 0$$
$$(2x + 2)(x + 5) = 0$$
$$x = -1 \text{ or } x = -5$$

Step 6: Check Your Answer

Let x = −1

$$2x^3 + 10x^2 - 2x - 10 = 0$$
$$2(-1)^3 + 10(-1)^2 - 2(-1) - 10 = 0$$
$$2(-1) + 10(1) - (-2) - 10 = 0$$
$$-2 + 10 + 2 - 10 = 0$$
$$0 = 0 \quad \textit{Correct!}$$

9 Turn the expression you created in step **8** into an equation by adding "= 0" to the end of the expression.

10 If the equation you created in step **9** is a quadratic equation, solve the equation. In this example, we solved the quadratic equation using the method shown on page 208.

Note: If the equation you created in step **9** is not a quadratic equation, see the tip at the top of this page. A quadratic equation is an equation whose highest exponent in the equation is two, such as x^2.

• In this example, −1 and −5 prove to be the other solutions to the equation.

11 To check your answer, place one of the solutions you found in step **10** into the original equation and solve the problem. If both sides of the equation are equal, you have correctly found one solution to the equation.

Note: In this example, you do not need to check the solution of 1 since you already confirmed this was a solution in step **4**.

12 Repeat step **11** to check the other solution you found.

• In this example, the solutions to the equation are 1, −1 and −5.

Solve Quadratic Inequalities

Inequalities are mathematical statements which compare different values. Unlike equations, such as x equals 4, inequalities make more general statements of value, such as x can be any number greater than 4.

A quadratic inequality is an inequality whose highest exponent is 2, such as $x^2 + 3x > 4$. Solving a quadratic inequality is quite similar to solving a quadratic equation, in that you can arrive at more than one answer. In fact, pretending the inequality is an equation is part of the process. After setting one side of the inequality to 0, you temporarily change the inequality symbol in the expression to an equals sign. Then you factor the equation and find the solutions for x.

The solutions for x usually split a number line into three segments, or ranges, on the number line—the range below the smallest number, the range above the largest number and the range between the two numbers. For example, if you found that $x = -2$ or $x = 3$, you would form three new inequalities to represent the ranges: $x < -2$, $-2 < x < 3$ and $x > 3$.

Step 1: Set One Side of the Inequality to 0

Solve the inequality $x^2 + 2x > 3$.

$$x^2 + 2x > 3$$
$$x^2 + 2x - 3 > 3 - 3$$
$$x^2 + 2x - 3 > 0$$

Step 2: Turn the Inequality into an Equation

$$x^2 + 2x - 3 > 0$$
$$x^2 + 2x - 3 = 0$$
$$(x + 3)(x - 1) = 0$$

$$x = -3$$
$$x = 1$$

---

1 Determine which numbers and/or variables you need to add or subtract to place all the variables and numbers on the left side of the inequality and make the right side of the inequality set to 0.

2 Add and subtract the numbers and/or variables you determined in step **1** on both sides of the inequality.

- In this example, subtract 3 from both sides of the inequality.

  Note: For more information on adding and subtracting numbers and variables in inequalities, see page 106.

3 Pretend the inequality symbol ($>$, $\geq$, $<$ or $\leq$) is an equals sign ($=$) to turn the inequality into an equation.

4 Solve the equation.

Note: In this example, you can use the method described on page 198 to solve the equation.

- In this example, the solutions to the equation are -3 and 1.

Tip

How do I know which inequality symbols to use when I form the ranges of possible numbers?

In step **6** below, if the original inequality includes the < or > symbol, as in $x^2 + 2x > 3$, you would write the inequalities that represent the ranges using the < and > symbols, as in $x < -3$, $-3 < x < 1$ and $x > 1$. If the original inequality includes the ≤ or ≥ symbol, as in $x^2 + 2x \geq 3$, you would write the inequalities using the ≤ and ≥ symbols, as in $x \leq -3$, $-3 \leq x \leq 1$ and $x \geq 1$.

Tip

What should I do if I create an equation that has no solutions?

Sometimes, when you turn an inequality into an equation, you will find that the equation has no solutions. In this case, either all numbers or no numbers are a solution to the original inequality. To determine which is the case, place any number, such as 0, into the inequality and solve the inequality. If the number you chose makes the inequality true, all numbers are solutions. If the number does not make the inequality true, there are no solutions. For instance, if you replace x with 0 in $x^2 + 5 > 0$, you end up with $5 > 0$. Since this inequality is true, all numbers are a solution to this inequality.

Step 3: Select Test Numbers

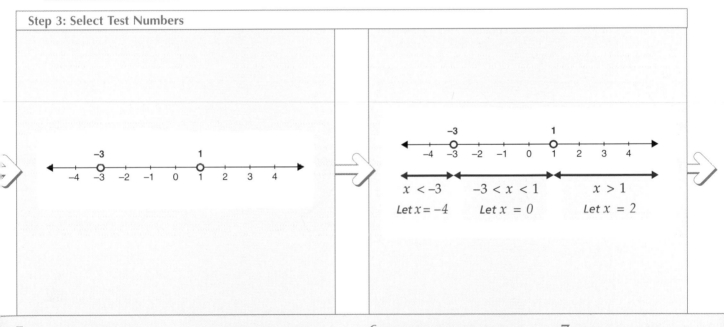

5 List the solutions to the equation in order, from smallest to largest.

- You may find it helpful to mark the solutions on a number line.

- In this example, −3 and 1 are marked on a number line.

Note: If the original inequality includes the symbol < or >, use an open circle (O) on the number line to indicate the number is not a solution. If the original inequality includes the symbol ≤ or ≥, use a filled-in circle (●) to indicate the number is a possible solution.

6 To determine if the solution to the inequality consists of numbers less than the smallest solution to the equation, numbers in between the two solutions, and/or numbers greater than the largest solution, write an inequality that describes each of these three possible solutions.

7 Select a number that is less than the smallest solution, in between the two solutions and greater than the largest solution.

CONTINUED

Solve Quadratic Inequalities *continued*

Once you have determined the three ranges of numbers for your inequality, pick a number from each range. These are test values that you will plug into the original inequality. If the test value from a given range makes the inequality true, all the values in that range are part of the solution to the inequality.

You do not need to test all three ranges. Testing just one of the test values is sufficient to tell you which ranges are solutions to the inequality. If the range that is between the two numbers is a solution to the inequality, the other two ranges are not solutions to the inequality. For example, if the range $-2 < x < 3$ is a solution, then $x < -2$ and $x > 3$ are not solutions.

Conversely, if one of the ranges above or below a number is a solution to the inequality, both ranges above and below the numbers will be solutions to the inequality, while the range between the two numbers is not a solution to the inequality. For example, if $x < -2$ is a solution, $x > 3$ is a solution as well, but $-2 < x < 3$ is not a solution to the inequality.

Step 4: Find the Solution

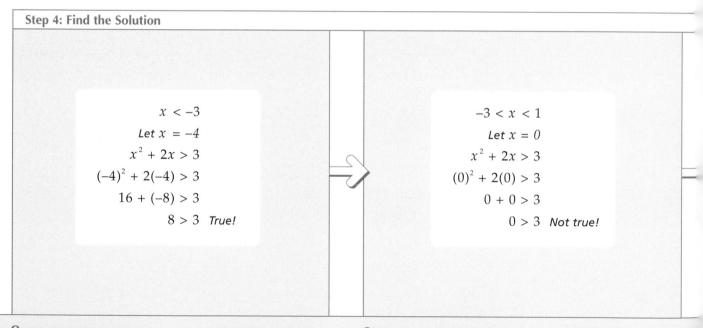

$$x < -3$$
$$\text{Let } x = -4$$
$$x^2 + 2x > 3$$
$$(-4)^2 + 2(-4) > 3$$
$$16 + (-8) > 3$$
$$8 > 3 \quad \textit{True!}$$

$$-3 < x < 1$$
$$\text{Let } x = 0$$
$$x^2 + 2x > 3$$
$$(0)^2 + 2(0) > 3$$
$$0 + 0 > 3$$
$$0 > 3 \quad \textit{Not true!}$$

8 Place one of the three numbers you selected into the original inequality and solve the inequality. If the number makes the inequality true, the range of numbers from which you selected the number is a solution to the inequality.

• In this example, the original inequality is true when you place -4 into the inequality, so $x < -3$ is a solution to the inequality.

9 Place one of the other two numbers you selected into the original inequality and solve the inequality. If the number makes the inequality true, the range of numbers from which you selected the number is a solution to the inequality.

• In this example, the original inequality is not true when you place 0 into the inequality, so $-3 < x < 1$ is not a solution to the inequality.

Practice

$a \cdot b - c$

Solve the following quadratic inequalities. You can check your answers on page 264.

1) $x^2 - 2x - 3 > 0$

2) $x^2 + 5 > 5$

3) $x^2 + 6x + 4 < 2x + 9$

4) $2x^2 + 3x + 2 \leq x^2 + 3x - 3$

5) $2x^2 - 6x + 9 \leq -2x^2 + 6x + 1$

6) $x^2 + 10x + 84 \leq -10x$

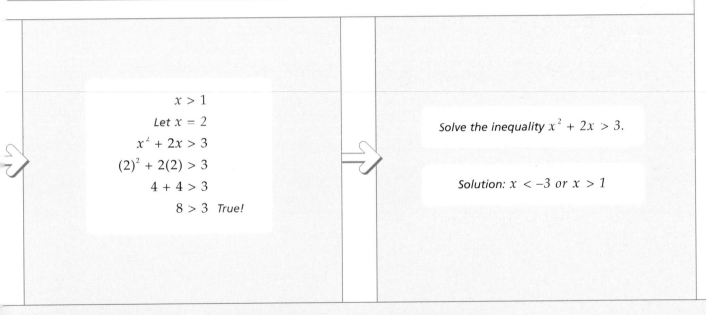

$x > 1$

Let $x = 2$

$x^2 + 2x > 3$

$(2)^2 + 2(2) > 3$

$4 + 4 > 3$

$8 > 3$ *True!*

Solve the inequality $x^2 + 2x > 3$.

Solution: $x < -3$ or $x > 1$

10 Place the last number you selected into the original inequality and solve the inequality. If the number makes the inequality true, the range of numbers from which you selected the number is a solution to the inequality.

- In this example, the original inequality is true when you place 2 into the inequality, so $x > 1$ is a solution to the inequality.

- The solution to the inequality $x^2 + 2x > 3$ is $x < -3$ or $x > 1$.

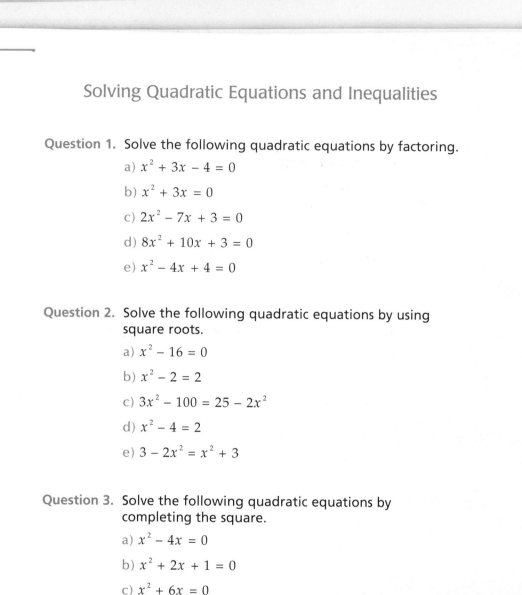

Solving Quadratic Equations and Inequalities

Question 1. Solve the following quadratic equations by factoring.

a) $x^2 + 3x - 4 = 0$

b) $x^2 + 3x = 0$

c) $2x^2 - 7x + 3 = 0$

d) $8x^2 + 10x + 3 = 0$

e) $x^2 - 4x + 4 = 0$

Question 2. Solve the following quadratic equations by using square roots.

a) $x^2 - 16 = 0$

b) $x^2 - 2 = 2$

c) $3x^2 - 100 = 25 - 2x^2$

d) $x^2 - 4 = 2$

e) $3 - 2x^2 = x^2 + 3$

Question 3. Solve the following quadratic equations by completing the square.

a) $x^2 - 4x = 0$

b) $x^2 + 2x + 1 = 0$

c) $x^2 + 6x = 0$

d) $3x^2 - 6x - 24 = 0$

e) $4x^2 - 2x = 2x + 3$

Question 4. Solve the following quadratic equations using the quadratic formula.

a) $x^2 - 3x + 2 = 0$

b) $x^2 + 5x = 2x$

c) $x^2 + 2x + 2 = -x^2 - x + 1$

d) $12x^2 + 5x = 2$

e) $2x^2 + 5x + 1 = 0$

Question 5. Find all of the solutions to the following equations.

a) $x^3 - x^2 + x - 1 = 0$

b) $x^4 - x^3 + x^2 - x = -2$

c) $2x^3 + 2x^2 = 2x + 2$

d) $3x^3 = 4x^2 + 13x + 6$

e) $x^4 - 10x^2 + 9 = 0$

Question 6. Solve the following quadratic inequalities.

a) $x^2 + x - 2 > 0$

b) $x^2 > 0$

c) $x^2 \leq 0$

d) $x^2 < 6 - x$

e) $x^2 + 5x \geq 15x$

f) $3x^2 - 4x + 1 \leq -3x^2 + 3x - 1$

You can check your answers on page 279.

Chapter 12

$y = 2(x-3)^2 + 2$

$y = 2(x+3)^2 - 2$

$y = 2(x-1)^2 - 1$

While the straight lines of linear equations are easy to graph, they are not the most exciting. Graphing quadratic equations, however, opens the door to the wonderful world of parabolas. Chapter 12 introduces these U-shaped curves and gives you the tools to graph them.

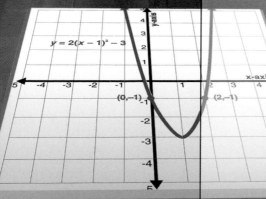

$y = 2(x-1)^2 - 3$

y-axis

x-axis

$(0,-1)$ $(2,-1)$

Graphing Quadratic Equations

In this Chapter...

Introduction to Parabolas

Remember being a kid and throwing that baseball as high as you could? The ball never landed in the same spot from which you tossed it because, whether you noticed it or not, the ball always followed a curve up and then fell back down on a similar curve. Those curves formed a symmetrical, U-shaped curve called a parabola.

When you graph a quadratic equation—an equation in which the highest exponent is two, such as x^2—the graph of the equation is a parabola. Parabolas open either downward or upward.

The lowest point in an upward-opening parabola or the highest point in a downward-opening parabola is called the vertex.

While quadratic equations can be expressed in the form $y = ax^2 + bx + c$, for the purposes of graphing, parabolas are generally rewritten as $y = a(x - h)^2 + k$. This is called the vertex form of an equation. In this form, the variable a determines how steep a parabola will be and whether it will open upward or downward. The variables h and k represent the position of the vertex (h,k) in the coordinate plane.

Parabola Basics

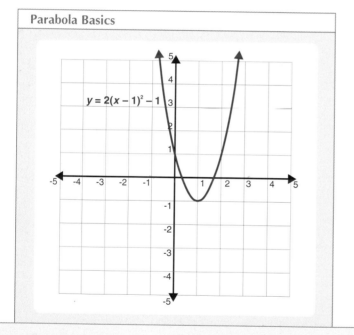

Steepness or Flatness of a Parabola

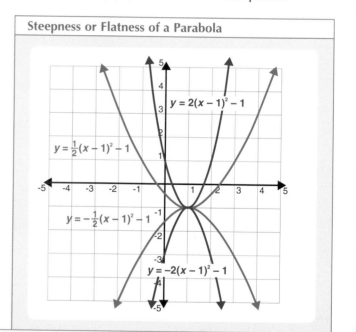

- When you graph a quadratic equation, you will end up with a U-shaped curve known as a parabola.

  Note: A quadratic equation is an equation whose highest exponent in the equation is two, such as x^2.

- The standard way to write a quadratic equation you want to graph is in the form $y = a(x - h)^2 + k$, where a, h and k are numbers. Writing a quadratic equation in this form provides information to help you graph the parabola.

  Note: To write a quadratic equation in the form $y = a(x - h)^2 + k$, see page 232.

- The number represented by a indicates the steepness or flatness of a parabola. The larger the number, whether the number is positive or negative, the steeper the parabola.

- The number represented by a also indicates if a parabola opens upward or downward. When a is a positive number, the parabola opens upward. When a is a negative number, the parabola opens downward.

Tip

Can the graph of a quadratic equation with just one term be a parabola?

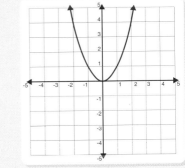

Yes. The graphs of even the simplest quadratic equations, which by definition must contain the x^2 variable, form a parabola. For example, a graph of the equation $y = x^2$ forms a parabola and can be written as $y = 1(x - 0)^2 + 0$ in the vertex form. In this equation, you will notice that $a = 1$, so the parabola opens upward, and because h and k both equal 0, the vertex is at the origin (0,0) in the coordinate plane.

Practice

$a + b = c$

Determine the vertex of each of the following parabolas and indicate if each parabola opens upward or downward. You can check your answers on page 264.

1) $y = (x - 1)^2$

2) $y = \frac{1}{2}(x - \frac{1}{3})^2 - 5$

3) $y = -3(x + 1)^2 + 2$

4) $y = 4x^2 + \frac{3}{10}$

5) $y = -x^2 + 2$

6) $y = -\frac{3}{4}(x - 2)^2 - 3$

Vertex of a Parabola

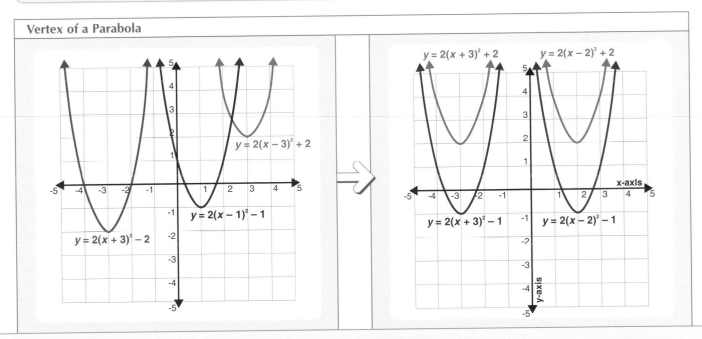

- The numbers represented by h and k, when written as (h,k), indicate the location of the vertex of a parabola, which is the lowest or highest point of a parabola.

Note: If a minus sign (−) appears after x in the equation, h in (h,k) is a positive number. If a plus sign (+) appears after x in the equation, h in (h,k) is a negative number.

- For example, in the equation $y = 2(x + 3)^2 - 2$, the vertex of the parabola is located at $(-3, -2)$.

- The number represented by h indicates how far left or right along the x-axis the vertex of a parabola is located from the origin. The origin is the location where the x-axis and y-axis intersect.

Note: When a minus sign (−) appears after x in the equation, the parabola moves to the right. When a plus sign (+) appears after x in the equation, the parabola moves to the left.

- The number represented by k indicates how far up or down along the y-axis the vertex of a parabola is located from the origin.

Note: Positive k numbers move the parabola up. Negative k numbers move the parabola down.

Write a Quadratic Equation in Vertex Form

Graphing a parabola is simple if you first change the quadratic equation from standard form ($y = ax^2 + bx + c$) to vertex form ($y = a(x - h)^2 + k$). The vertex form of a quadratic equation gives you the direction that the parabola opens, as well as the coordinates of the vertex—the highest point of a parabola that opens downward or lowest point of a parabola that opens upward.

To convert a quadratic equation to vertex form, you must determine the values of h and k—the

value of a is the same in both equations. To determine the value of h, you use a simple formula ($h = \frac{-b}{2a}$) that uses the standard form values of a and b. To determine the value of k, go back to the standard form of the equation. Replace the variable y with the variable k and replace the x variables with the value of h. You can then solve for the variable k in the equation. When you have all three values—a, h and k—you can write the equation out in vertex form.

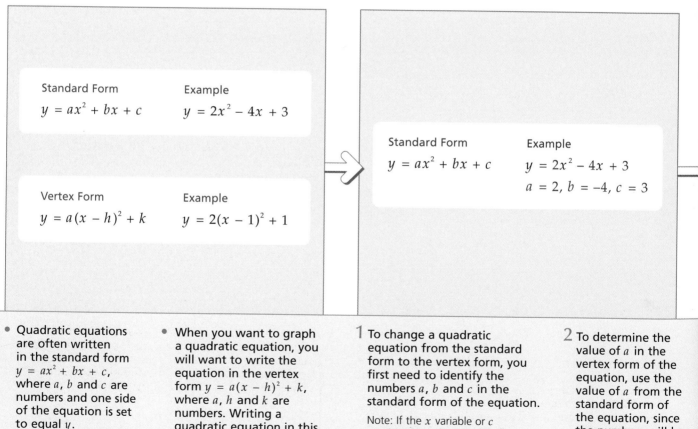

Standard Form

$y = ax^2 + bx + c$

Example

$y = 2x^2 - 4x + 3$

Vertex Form

$y = a(x - h)^2 + k$

Example

$y = 2(x - 1)^2 + 1$

Standard Form

$y = ax^2 + bx + c$

Example

$y = 2x^2 - 4x + 3$

$a = 2, b = -4, c = 3$

- Quadratic equations are often written in the standard form $y = ax^2 + bx + c$, where a, b and c are numbers and one side of the equation is set to equal y.

  Note: A quadratic equation is an equation whose highest exponent in the equation is two, such as x^2.

- When you want to graph a quadratic equation, you will want to write the equation in the vertex form $y = a(x - h)^2 + k$, where a, h and k are numbers. Writing a quadratic equation in this form provides information to help you graph the equation.

1 To change a quadratic equation from the standard form to the vertex form, you first need to identify the numbers a, b and c in the standard form of the equation.

  Note: If the x variable or c number does not appear in the equation, assume the number is 0. For example, the equation $y = 2x^2 - 4x$ is the same as $y = 2x^2 - 4x + 0$.

2 To determine the value of a in the vertex form of the equation, use the value of a from the standard form of the equation, since the numbers will be the same.

- In this example, a equals 2.

Tip

How do I change an equation from vertex form to the standard form of a quadratic equation?

If you multiply and then simplify the terms of an equation in vertex form, you will arrive at the standard form of the quadratic equation. This is a great way to check your answer after converting an equation from standard form to vertex form. For information on multiplying polynomials, see page 154.

$$y = 2(x - 1)^2 + 1$$
$$y = 2(x - 1)(x - 1) + 1$$
$$y = 2(x^2 - 2x + 1) + 1$$
$$y = 2x^2 - 4x + 2 + 1$$
$$y = 2x^2 - 4x + 3$$

Practice

$a + b = c$

Write the following quadratic equations in vertex form. You can check your answers on page 264.

1) $y = x^2 - 2x + 3$

2) $y = -x^2 + 4x - 1$

3) $y = -2x^2 - 4x - 3$

4) $y = x^2 - 6x + 7$

5) $y = x^2 + 3$

6) $y = x^2 + 10x + 15$

$$h = \frac{-b}{2a}$$

$$= \frac{-(-4)}{2 \times 2} = \frac{4}{4} = 1$$

$$k = 2h^2 - 4h + 3$$
$$= 2(1)^2 - 4(1) + 3$$
$$= 2 - 4 + 3$$
$$= 1$$

$$y = a(x - h)^2 + k$$
$$y = 2(x - 1)^2 + 1$$

3 To determine the value of h in the vertex form of the equation, use the numbers a and b from the standard form of the equation in the formula $h = \frac{-b}{2a}$.

- In this example, h equals 1.

4 To determine the value of k, use the standard form of the equation and replace the variable y with the variable k and replace the x variables with the value of h you determined in step **3**. Then solve for k in the equation.

- In this example, k equals 1.

5 In the vertex form of the equation, replace the numbers you determined for a, h and k.

- You have finished changing the quadratic equation from the standard form to the vertex form.

233

Graph a Parabola

When you graph a quadratic equation in the form of $y = ax^2 + bx + c$ in the coordinate plane, you will always end up with a U-shaped curve known as a parabola. For information on the coordinate plane, see page 80.

When you want to graph a quadratic equation, it is useful to have the equation written in vertex form $(y = a(x - h)^2 + k)$. The vertex form tells you whether the parabola opens upward or downward and provides you with the coordinates of the parabola's vertex—the highest or lowest point of the parabola. In an equation in vertex form, if the value of a is positive, the parabola opens upward, while if the value of a is negative, the parabola opens downward. The values h and k represent the coordinates (h,k) of the parabola's vertex.

After you plot the vertex, plotting one point on either side of the vertex is sufficient to draw the rest of the curve. If you want to make sure you did not make a mistake, you can plot more points.

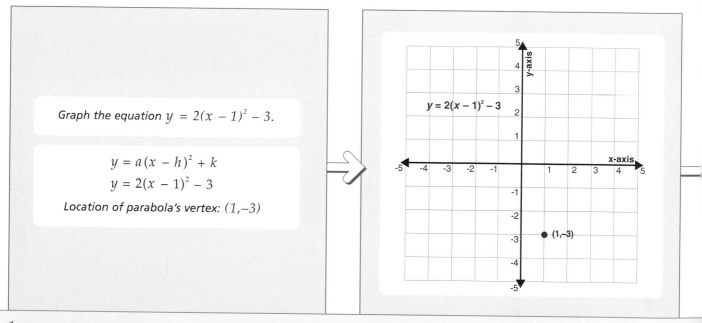

Graph the equation $y = 2(x - 1)^2 - 3$.

$$y = a(x - h)^2 + k$$
$$y = 2(x - 1)^2 - 3$$

Location of parabola's vertex: $(1,-3)$

1 To determine the location of the vertex of the parabola, which is the lowest or highest point of the parabola, look at the h and k numbers in the equation.

Note: If a minus sign (−) appears after x in the equation, the value of h in (h,k) is a positive number. If a plus sign (+) appears after x in the equation, the value of h in (h,k) is a negative number.

2 Write the h and k numbers as an ordered pair in the form (h,k). An ordered pair is two numbers, written as (x,y), which gives the location of a point in the coordinate plane.

3 Plot the point for the parabola's vertex in the coordinate plane.

Note: To plot points in a coordinate plane, see page 82.

4 To determine if the parabola will open upward or downward, look at the number represented by a. When the value of a is a positive number, the parabola will open upward. When the value of a is a negative number, the parabola will open downward.

● In the equation $y = 2(x - 1)^2 - 3$, the parabola will open upward since the value of a, or 2, is a positive number.

234

Tip

Can graphing a parabola help me solve a quadratic equation?

Yes. Look at where a parabola crosses the x-axis. The values for x at those points will give you the solutions to the quadratic equation. If the parabola crosses the x-axis at two points, the equation has two solutions. If the parabola touches the x-axis at just one point, the equation has one solution. If the parabola does not cross the x-axis, the equation has no real solutions.

Practice

$a+b=c$

Graph the following parabolas. You can check your graphs on page 265.

1) $y = -(x - 1)^2 + 3$

2) $y = (x + 2)^2 - 4$

3) $y = x^2 - 1$

4) $y = -x^2 + 3$

5) $y = (x - 2)^2 - 1$

6) $y = -2(x + 3)^2$

Let $x = 0$

$y = 2(x - 1)^2 - 3$

$y = 2(0 - 1)^2 - 3$

$y = 2 - 3$

$y = -1$

Ordered pair $(x,y) = (0,-1)$

Let $x = 2$

$y = 2(x - 1)^2 - 3$

$y = 2(2 - 1)^2 - 3$

$y = 2 - 3$

$y = -1$

Ordered pair $(x,y) = (2,-1)$

$y = 2(x - 1)^2 - 3$

---

5 Choose a random number for the x variable. For example, let x equal 0.

6 Place the number you selected into the equation to determine the value of the y variable. Then solve for y in the equation.

7 Write the x and y value together as an ordered pair in the form (x,y).

8 Repeat steps **5** to **7** to determine another ordered pair. The ordered pair should be located on the other side of the parabola's vertex so you have one point on either side of the parabola's vertex.

9 Plot the two points in a coordinate plane.

10 Connect the points to draw a smooth curve. Draw an arrow at each end of the curve to show that the parabola extends forever.

- The parabola shows all the possible solutions to the equation.

235

Graphing Quadratic Equations

Question 1. Determine the vertex of the following parabolas and whether the parabolas open upward or downward.

a) $y = (x - 1)^2 + 2$

b) $y = -(x + 6)^2 - 3$

c) $y = 4(x + 4)^2 + \dfrac{1}{2}$

d) $y = -\dfrac{1}{2}\left(x - \dfrac{1}{3}\right)^2$

e) $y = 200(x - 50)^2 - 75$

Question 2. Write the following quadratic equations in vertex form. Determine the vertex of each parabola and whether the parabola opens upward or downward.

a) $y = x^2 - 2x + 1$

b) $y = x^2 + 6x + 5$

c) $y = -x^2 - 2x + 2$

d) $y = -4x^2 + 4x + 3$

e) $y = 25x^2 + 20x + 16$

Question 3. Graph the following parabolas.

a) $y = (x + 1)^2 - 2$

b) $y = -(x - 3)^2 + 1$

c) $y = 2x^2 - 8x + 5$

d) $y = x^2 + 4x$

e) $y = -2x^2 - 12x - 14$

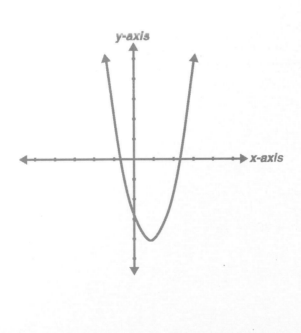

*You can check your answers
on pages 280-281.*

Chapter 13

This chapter is a great resource for quickly accessing definitions for algebra terms you may not be familiar with. When you need the explanation of an algebra term, you can easily refer to the comprehensive glossary in this chapter.

Glossary

In this Chapter...

Glossary

Glossary

A

Absolute Value

The absolute value of a number is always the positive value of the number, whether the number is positive or negative. For example, the absolute value of 5 and –5 is 5. An absolute value is indicated by two thin vertical lines around a number, such as $|5|$.

B

Base

The base in an exponential expression is the number or variable being multiplied by itself. For example, in the expression 6^2, the base is 6, while the exponent is 2.

Binomial

A binomial is a polynomial with two terms, such as $5x + 7$.

C

Coefficient

A coefficient is a number in front of a variable. For example, 5 is the coefficient in the expression $5x$.

Common Denominator

A common denominator is a denominator, or bottom number, shared by a group of fractions. For example, $\frac{1}{4}$ and $\frac{3}{4}$ have a common denominator of 4. To add and subtract fractions, the fractions must have a common denominator.

Composite Number

A composite number is a number that you can divide evenly by itself, the number 1 and one or more other numbers. For example, 4, 6 and 8 are composite numbers.

Compound Inequality

A compound inequality is the combination of two inequalities, which are written together. For example, $2 < x < 8$ is a compound inequality.

Coordinate

A coordinate is one part of an ordered pair (x,y). The x coordinate tells you how far along the x-axis a point is located in a coordinate plane, while the y coordinate tells you how far along the y-axis a point is located in a coordinate plane.

Coordinate Plane

A coordinate plane is a grid used to graph equations. The coordinate plane contains a horizontal line, called the x-axis, and a vertical line, called the y-axis, which intersect at a point called the origin.

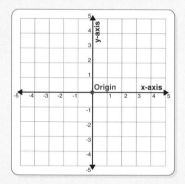

Cube Root

The cube root of a number is a number you can multiply by itself three times to equal the given number. For example, the cube root of 8, written as $\sqrt[3]{8}$, is 2, since $2 \times 2 \times 2$ equals 8.

D

Cubed

When a number or variable is said to be cubed, it means the number or variable is multiplied by itself three times. For example, b^3 can be read as "b cubed," and means $b \times b \times b$.

Degree

The degree of an expression refers to the highest exponent in the expression. For example, $x^2 + x + 5$ has a degree of 2. If an expression contains more than one variable, such as x and y, the degree is determined by adding the exponents of each term together. The term with the highest total of exponents determines the degree of the expression. For example, $x^2y^3 + xy$ has a degree of 5.

Denominator

A denominator is the bottom number in a fraction. For example, in the fraction $\frac{3}{4}$, the denominator is 4.

Difference

The difference is the answer to a subtraction problem.

Distributive Property

The distributive property allows you to eliminate a set of parentheses ( ) by multiplying each number and variable within the parentheses by a number or variable outside the parentheses. For example, $3(a + b)$ equals $3a + 3b$.

Dividend

The dividend is the number being divided by another number in a division problem. For example, in the division problem $10 \div 5$, 10 is the dividend.

Divisor

The divisor is the number being divided into another number in a division problem. For example, in the division problem $10 \div 5$, 5 is the divisor.

E

Equation

An equation is a mathematical statement containing an equals sign (=), indicating that the two sides of the equation are equal.

Even Number

An even number is a number that you can divide evenly by 2. For example, 2, 4, 6 and 8 are even numbers.

Exponent

An exponent, or power, appears as a small number above and to the right of a number or variable, such as in 2^3. An exponent indicates the number of times a number or variable is multiplied by itself. For example, 2^3 equals $2 \times 2 \times 2$.

Expression

An expression is a mathematical statement that does not contain an equals sign (=). An expression can contain numbers, variables and/or operators such as +, −, × and ÷. For example, $x + y$ and $5x - 2y$ are expressions.

F

Factor

Factors are numbers that you can multiply together to end up with a specific number. For example, 3 and 4 are factors of 12. Factors can also include variables. When you factor an expression in algebra, you break the expression into pieces, called factors, that you can multiply together to give you the original expression.

Glossary

Formula

A formula is a statement expressing a general mathematical truth and can be used to solve or reorganize mathematical problems. For example, you can use the following formula to find the slope, or steepness, of a line.

$$m = \frac{y_2 - y_1}{x_2 - x_1}$$

Fraction

A fraction, such as $\frac{1}{2}$ or $\frac{3}{4}$, is a division problem written with a fraction bar (—) instead of a division sign ($\div$). For example, you can write $3 \div 4$ as $\frac{3}{4}$. A fraction has two parts—the top number in a fraction is called the numerator and the bottom number in a fraction is called the denominator.

Graph

A graph is a visual representation of all the solutions to an equation or inequality. You graph a line or a parabola, which is a U-shaped curve, using points plotted in a coordinate plane.

Greatest Common Factor

The greatest common factor (GCF) is the largest term that divides evenly into each term in an expression. For example, in the expression $5a + 10b + 15c$, 5 is the greatest common factor.

Grouping Symbols

Grouping symbols include parentheses ( ), brackets [ ] and braces { }. You should always work with numbers and variables inside grouping symbols first. For example, in the problem $5 \times (2 + 3)$, you should add $2 + 3$ and then multiply the result by 5.

Improper Fraction

An improper fraction is a fraction in which the numerator, or top number, is larger than the denominator, or bottom number. For example, $\frac{3}{2}$ and $\frac{4}{3}$ are improper fractions. An improper fraction can also be written as a mixed number. For example, $\frac{3}{2}$ can be written as $1\frac{1}{2}$.

Index

The index of a radical is the small number just in front of a radical sign. For example, in the radical $\sqrt[3]{8}$, 3 is the index. The index tells you which root you must find. For example, the radical $\sqrt[3]{8}$ asks you to find the third root of 8, which is the number you can multiply by itself three times to equal 8. If an index does not appear in front of a radical sign, as in $\sqrt{16}$, assume the index is 2.

Inequality

An inequality is a mathematical statement in which one side is less than, greater than, or possibly equal to the other side. Inequalities use four different symbols—less than (<), less than or equal to ($\leq$), greater than (>) and greater than or equal to ($\geq$). For example, the inequality $10 < 20$ states that 10 is less than 20.

Integer

An integer is a whole number or a whole number with a negative sign (–) in front of the number. For example, –3, –2, –1, 0, 1, 2 and 3 are integers.

Intercept

An intercept is a point where a line crosses the x-axis or y-axis in a coordinate plane. An x-intercept is the point where a line crosses the x-axis, while a y-intercept is the point where a line crosses the y-axis.

L

Irrational Number

An irrational number is a number that has decimal values that continue forever without a repeating pattern. Irrational numbers do not include integers or fractions. The most well-known irrational number is pi (π), which is equal to 3.1415926…

Least Common Denominator

The least common denominator is the smallest denominator, or bottom number, shared by a group of fractions. For example, $\frac{1}{2}$ and $\frac{3}{4}$ have a least common denominator of 4.

Like Terms

Terms that contain exactly matching variables are called like terms. For example, x, $2x$ and $3x$ are all like terms.

Linear Equation

A linear equation is an equation whose highest exponent of a variable in the equation is one, such as in $x = 10$.

M

Matrix

A matrix is a collection of numbers, called elements, which are arranged in horizontal rows and vertical columns. The collection of numbers is surrounded by brackets. Matrices is the term used to indicate more than one matrix.

$$\begin{bmatrix} 1 & 7 & 5 \\ 2 & 6 & 4 \\ 8 & 3 & 9 \end{bmatrix}$$

Mixed Number

A mixed number consists of a whole number followed by a fraction. For example, $3\frac{1}{2}$ is a mixed number. Mixed numbers can be written as improper fractions. For example, $3\frac{1}{2}$ can be written as $\frac{7}{2}$.

Monomial

A monomial is a polynomial with only one term, such as $5x$.

N

Natural Number

Natural numbers, also called the counting numbers, include the numbers 1, 2, 3, 4, 5 and so on.

Negative Number

A negative number is a number that is less than 0 and is written with a minus sign (–) in front of the number, such as –4. Negative numbers become smaller the farther they are from zero. For example, –10 is smaller than –5.

Number Line

A number line consists of a line on which numbers are assigned to equally spaced points. A number line can be used to graph an inequality, such as $x > 2$, to visually indicate all the possible solutions to the inequality.

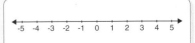

Glossary

Numerator

The numerator is the top number in a fraction. For example, in the fraction $\frac{3}{4}$, the numerator is 3.

Odd Number

An odd number is a number that you cannot divide evenly by 2. When you divide an odd number by 2, you get a left-over value, known as a remainder.

Order of a Matrix

The order of a matrix refers to the size and shape of a matrix and indicates the number of rows and columns in a matrix. The order of a matrix is written as the number of rows, followed by an x and then the number of columns, such as 3 × 4.

Order of Operations

The Order of Operations is a specific order that you should use when solving math problems. The Order of Operations specifies that you should work first with numbers in parentheses, then calculate exponents, then multiply and divide, and then add and subtract.

Ordered Pair

An ordered pair is two numbers, called coordinates, written as (x,y), that gives the location of a point in a coordinate plane. The first number in an ordered pair, known as the x coordinate, tells you how far along the x-axis a point is located. The second number in an ordered pair, known as the y coordinate, tells you how far along the y-axis a point is located.

Origin

The origin in a coordinate plane is the location where the x-axis and y-axis intersect. The ordered pair representing the origin is (0,0).

Parabola

A parabola is a symmetrical, U-shaped curve that results when you graph a quadratic equation.

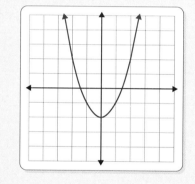

Parallel

Parallel lines never intersect and have the same slope, which always keeps the lines the same distance apart from one another.

Perpendicular

Perpendicular lines intersect one another at right, or 90-degree, angles. Perpendicular lines have slopes that are negative reciprocals of each other. For example, $-\frac{1}{2}$ is the negative reciprocal of 2.

Plot

To plot a point in a coordinate plane means to mark the location of the point in the coordinate plane.

Point-Slope Equation

The point-slope equation allows you to create the equation of a line by using the slope of the line, represented by m, and one of the points on the line, represented by (x_1, y_1). The point-slope equation is written as $y - y_1 = m(x - x_1)$.

Polynomial

A polynomial consists of one or more terms, which can be a combination of numbers and/or variables, that are added together or subtracted from one another. For example, $2x + 3y$ is a polynomial.

Positive Number

A positive number is a number that is greater than 0 and can be written with or without a positive sign (+). For example, +4 can also be written as 4. Positive numbers become larger the farther they are from zero. For example, 60 is larger than 20.

Prime Number

A prime number is a positive number that you can only evenly divide by itself and the number 1. For example, 2, 3, 5, 7 and 11 are examples of prime numbers. The number 1 is not considered a prime number.

Product

The product is the answer to a multiplication problem.

Q

Quadrant

A coordinate plane is divided into four quadrants by the x-axis and y-axis. The four quadrants are labelled with the Roman numerals I, II, III and IV, starting with the top right quadrant and moving counterclockwise.

Quadratic Equation

A quadratic equation is an equation whose highest exponent of a variable in the equation is two, such as in $x^2 + x + 16 = 0$.

Quadratic Formula

The quadratic formula can be used to solve any quadratic equation. The quadratic formula is stated as:

$$x = \frac{-b \pm \sqrt{b^2 - 4ac}}{2a}$$

Quotient

The quotient is the answer to a division problem.

R

Radical

A radical is a symbol ($\sqrt{}$) that tells you to find the root of a number. In algebra, you will commonly find the square root of numbers. For example, $\sqrt{25}$ equals 5.

Radicand

The radicand is the number that appears under a radical sign, such as 16 in the expression $\sqrt{16}$.

Glossary

Rational Number

A rational number is an integer or a fraction. A rational number is a number that you can write with decimal values that either end or have a pattern that repeats forever. The numbers 3, −5, $\frac{1}{2}$, 9.25 and 0.2424... are rational numbers.

Real Number

A real number can be a natural number, whole number, integer, rational number or irrational number.

Reciprocal

Every number, except for zero, has a reciprocal number. When a number is multiplied by its reciprocal, the answer will always be 1. To find the reciprocal of a fraction, flip the top and bottom numbers in the fraction. For instance, the reciprocal of $\frac{2}{3}$ is $\frac{3}{2}$. To find the reciprocal of a whole number, write the number as 1 divided by the number. For instance, the reciprocal of 3 is $\frac{1}{3}$.

Remainder

The remainder is the left-over value in the result of a division problem when a number does not evenly divide into another number.

S

Scalar

A scalar is a number outside a matrix, by which all the elements in a matrix are multiplied.

Scientific Notation

Scientific notation is a shorthand method that uses exponents for writing very large or very small numbers in a more compact form. For example, the number 325000 can be rewritten as 3.25×10^5 in scientific notation.

Simplify

To simplify an expression, you combine all the terms that can be combined. For example, $2x + 4x$ equals $6x$.

Slope

The slope of a line indicates the steepness and direction of a line.

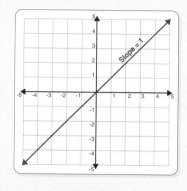

Slope-Intercept Form

When the equation of a line is written in slope-intercept form, you can immediately identify the slope of the line and the y-intercept of the line. The slope-intercept form of a line is written as $y = mx + b$, where m is the slope of the line and b is the y-intercept.

Solve

When you are asked to solve a problem, you need to find the answer to the problem. For example, when you solve for x in the equation $2x - 6 = 0$, you will determine that x equals 3.

Square Root

The square root of a number is a number you can multiply by itself to equal the given number. For example, the square root of 16, written as $\sqrt{16}$, is 4, since 4×4 equals 16.

Squared

When a number or variable is said to be squared, it means the number or variable is multiplied by itself. For example, 7^2 can be read as "7 squared," and means 7×7.

Sum

The sum is the answer to an addition problem.

Synthetic Division

Synthetic division is a shortcut method of dividing polynomials. This method can be used only when you are dividing by a polynomial with only two terms in the form $x - c$, where c is a number.

System of Equations

A system of equations is a group of two or more equations. You are often asked to find the value of each variable that solves both equations.

Terms

Terms are a combination of numbers and/or variables, which are separated by addition (+) or subtraction (–) signs. For example, the expression xy contains one term, while the expression $2x + 3y$ contains two terms.

Trinomial

A trinomial is a polynomial with three terms, such as $2x^2 + 5x + 7$.

Undefined

An answer of undefined in algebra means that no answer exists for a problem or expression. For example, the expression 0^0 is said to be undefined.

Variable

A variable is a letter, such as x or y, which represents an unknown number. For example, if x represents Emily's age, then $x + 5$ represents the age of Emily's sister who is five years older.

Vertex

The vertex of a parabola is the lowest or highest point of a parabola.

Vertex Form

The vertex form of a quadratic equation provides information to help you graph the equation. The vertex form of a quadratic equation is written as $y = a(x - h)^2 + k$, where a, h and k are numbers.

Whole Number

Whole numbers include the numbers 0, 1, 2, 3, 4, 5 and so on.

x-axis

The x-axis is a horizontal line that divides a coordinate plane. On the x-axis, the numbers to the right of the origin are positive and the numbers to the left of the origin are negative.

y-axis

The y-axis is a vertical line that divides a coordinate plane. On the y-axis, the numbers above the origin are positive and the numbers below the origin are negative.

Chapter 14

Are you ready to check the solutions for all of the problems you worked on throughout this book? This chapter reveals the answers to all of those questions you worked so hard on solving.

Solutions

In this Chapter...

Practice Solutions

Test Your Skills Solutions

Practice Solutions

Chapter 1 — Algebra Basics

Classifying Numbers (page 17)

1) False. 5 is not a composite number.
2) True.
3) False. 20 is not an odd number.
4) False. −7 is not a prime number.
5) True.
6) False. 43 is not a composite number.

Classifying Numbers (page 19)

1) integer, rational, real
2) irrational, real
3) integer, rational, real
4) irrational, real
5) rational, real
6) whole, integer, rational, real

Add and Subtract Numbers (page 21)

1) 14
2) −2
3) 3
4) 5
5) 8
6) −6

Multiply and Divide Numbers (page 23)

1) 5
2) 6
3) 5

4) $-2x$
5) 12
6) −2

Order of Operations (page 27)

1) $3 - (-3) = 6$
2) $2 \times 3 + (-2)^3 = 6 + (-8) = -2$
3) $\dfrac{(-2)}{2} = -1$
4) $2 - 15 = -13$
5) $2 \times 3 = 6$
6) $2^2 = 4$

The Distributive Property (page 31)

1) $2 \times 3 - 2 \times 5 = -4$
2) $3 \times 7 - 3 \times 4 = 9$
3) $5 \times 1 + 5 \times 2 = 15$
4) $(-1) \times 10 + (-1) \times 4 = -14$
5) $(-2) \times 3 - (-2) \times 2 = -2$
6) $(-5) \times (-1) - (-5) \times 4 = 25$

The Inverse Property (page 33)

1) −5
2) −7
3) 17
4) a
5) $-\dfrac{3}{4}$
6) 1

a) $\dfrac{1}{9}$
b) $\dfrac{4}{5}$
c) $\dfrac{1}{x}$
d) $\dfrac{b}{a}$
e) $-\dfrac{1}{8}$
f) $-\dfrac{4}{3}$

Add and Subtract Variables (page 37)

1) $-3x^3 + x^2 + 5x$
2) $2x + 7$
3) $2x^5 + x^4 + 3x$
4) $2x$
5) 0
6) $-3x + 5$

Multiply and Divide Variables (page 39)

1) $12x^7$ a) $2x$
2) -8 b) $5y^3$
3) $24x^7y$ c) $4y^2$
4) $6x^2y^2$ d) $3x^2y^2$
5) $-10z^{-3}$ e) $2x$
6) $6x^{-1}y^7 = \dfrac{6y^7}{x}$ f) $2a^2b^3$

Chapter 2 – Working With Fractions and Exponents

Convert Improper Fractions to Mixed Numbers (page 47)

1) $1\frac{1}{4}$ a) $\frac{5}{2}$
2) $3\frac{1}{2}$ b) $\frac{31}{3}$
3) 6 c) $\frac{22}{5}$
4) $3\frac{2}{3}$ d) $\frac{10}{7}$
5) $2\frac{1}{2}$ e) $\frac{23}{9}$
6) 3 f) $\frac{107}{10}$

Multiply and Divide Fractions (page 49)

1) $\dfrac{3}{8}$
2) $\dfrac{14}{3}$
3) $\dfrac{2}{5} \times \dfrac{1}{4} = \dfrac{2}{20} = \dfrac{1}{10}$
4) $\dfrac{3}{5} \times \dfrac{7}{2} = \dfrac{21}{10}$
5) $\dfrac{1 \times 5 \times 3}{2 \times 6 \times 4} = \dfrac{15}{48} = \dfrac{5}{16}$
6) $\dfrac{1}{2} \times \dfrac{5}{6} \times \dfrac{4}{3} = \dfrac{1 \times 5 \times 4}{2 \times 6 \times 3} = \dfrac{20}{36} = \dfrac{5}{9}$

Find the Least Common Denominator (page 51)

1) $\dfrac{3}{6}, \dfrac{2}{6}$
2) $\dfrac{5}{10}, \dfrac{3}{10}$
3) $\dfrac{20}{35}, \dfrac{14}{35}$
4) $\dfrac{6}{3}, \dfrac{1}{3}$
5) $\dfrac{9}{12}, \dfrac{8}{12}$
6) $\dfrac{16}{100}, \dfrac{25}{100}$

Exponent Basics (page 55)

1) $2 \times 2 \times 2 \times 2 = 16$
2) $3 \times 3 \times 5 \times 5 \times 5 = 9 \times 125 = 1125$
3) $(-3) \times (-3) \times (-3) = -27$
4) $-(4 \times 4) = -16$
5) $(3 \times 7) \times (3 \times 7) = 441$
6) $7 \times 7 \times 7 \times 7 \times 7 = 16807$

Practice Solutions

Chapter 2 – Working With Fractions and Exponents *continued*

Rules for Exponents (page 57)

1) $4^{2+3} = 4^5$
2) $x^{11-4} = x^7$
3) $2^{3+2-5} = 2^0 = 1$
4) $3^3 \times 3^8 \times 4^3 \times 4^5 = 3^{3+8} \times 4^{3+5} = 3^{11}4^8$
5) $y^{2+7-4} = y^5$
6) $10^{4-2+10} = 10^{12}$

Rules for Exponents (page 59)

1) $2^{3 \times 5} = 2^{15}$
2) $3^{(-2)5} = 3^{-10}$
3) $3^2 5^2$
4) $\dfrac{7^3}{4^3}$
5) $\dfrac{5}{9}$
6) $\left(\dfrac{4}{3}\right)^2 = \dfrac{4^2}{3^2}$

Chapter 3 – Solving Basic Equations

Solve an Equation with Addition and Subtraction (page 66)

1) $x = 2$
2) $x = 13$
3) $x = 9$
4) $x = 0$
5) $x = 10$
6) $x = 5$

Solve an Equation with Multiplication and Division (page 67)

1) $x = 2$
2) $x = 4$
3) $x = -3$
4) $x = -10$
5) $x = 21$
6) $x = \dfrac{15}{8}$

Solve an Equation in Several Steps (page 69)

1) $x = 2$
2) $x = 4$
3) $x = 15$
4) $x = -1$
5) $x = -2$
6) $x = 2$

Solve for a Variable in an Equation with Multiple Variables (page 71)

1) $x = \dfrac{y}{2} + 1$
2) $x = 2 - 3y$
3) $x = 3y + 7$
4) $y = -2x + 7$
5) $y = 3x - 2$
6) $y = 1$

Solve an Equation with Absolute Values (page 75)

1) $x = 5$ or $x = -5$
2) $x = 4$ or $x = -1$

3) $x = 5$ or $x = -3$

4) $x = 0$ or $x = \dfrac{8}{3}$

5) $x = 3$

6) No solution since an absolute value cannot equal a negative number ($|2x+6| = -1$).

Chapter 4 – Graphing Linear Equations

Introduction to the Coordinate Plane (page 83, practice 1)

A) $(-4,3)$

B) $(-3,1)$

C) $(-2,-3)$

D) $(3,3)$

E) $(1,-2)$

F) $(3,-1)$

Introduction to the Coordinate Plane (page 83, practice 2)

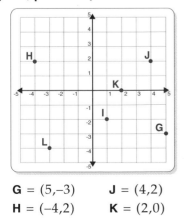

G = $(5,-3)$	**J** = $(4,2)$
H = $(-4,2)$	**K** = $(2,0)$
I = $(1,-2)$	**L** = $(-3,-4)$

Graph a Line (page 85)

1)

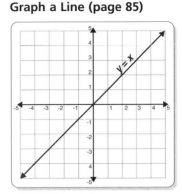

2)

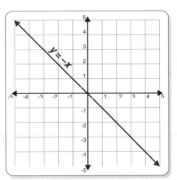

3)

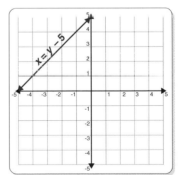

4)

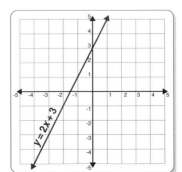

253

Chapter 4 – Graphing Linear Equations *continued*

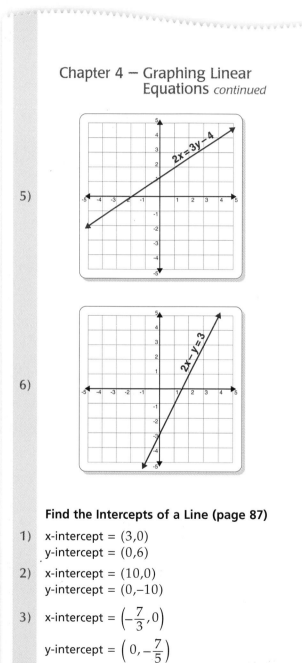

5)

6)

Find the Intercepts of a Line (page 87)

1) x-intercept = $(3,0)$
   y-intercept = $(0,6)$

2) x-intercept = $(10,0)$
   y-intercept = $(0,-10)$

3) x-intercept = $\left(-\frac{7}{3},0\right)$

   y-intercept = $\left(0,-\frac{7}{5}\right)$

4) x-intercept = $(5,0)$
   y-intercept = $(0,-5)$

5) no x-intercept
   y-intercept = $(0,5)$

6) x-intercept = $(-6,0)$
   no y-intercept

Find the Slope of a Line (page 89)

1) slope = $\frac{7}{3}$

2) slope = $-\frac{1}{2}$

3) slope = 2

4) slope = -2

5) slope = 1

6) slope = 0

Write an Equation in Slope-Intercept Form (page 91)

1) $y = x + 8$; slope = 1;
   y-intercept = $(0,8)$

2) $y = -3x - 5$; slope = -3;
   y-intercept = $(0,-5)$

3) $y = \frac{x}{3} - \frac{4}{3}$; slope = $\frac{1}{3}$;

   y-intercept = $\left(0,-\frac{4}{3}\right)$

4) $y = -x + 4$; slope = -1;
   y-intercept = $(0,4)$

5) $y = 0$; slope = 0;
   y-intercept = $(0,0)$

6) $y = \frac{x}{5} + \frac{1}{5}$; slope = $\frac{1}{5}$;

   y-intercept = $\left(0,\frac{1}{5}\right)$

Graph a Line in Slope-Intercept Form (page 93)

1)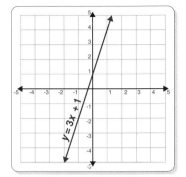

2)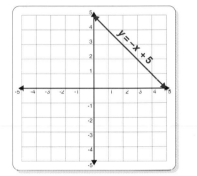

3)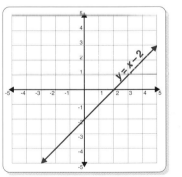

4)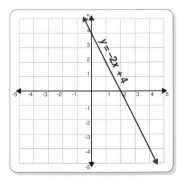

5)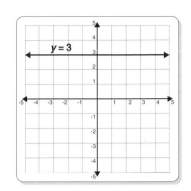

6)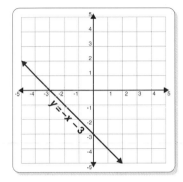

Write an Equation in Point-Slope Form (page 95)

1) $y - 3 = 2(x - 1)$

2) $y - 4 = -3x$

3) $y + 2 = -(x - 1)$

4) $y - 7 = 5(x + 3)$

5) $y - 1 = 0$

6) $y + 6 = 12(x + 3)$

Parallel and Perpendicular Lines (page 97)

1) parallel

2) perpendicular

3) perpendicular

4) parallel

5) parallel

6) neither parallel nor perpendicular

Practice Solutions

Chapter 4 – Graphing Linear Equations *continued*

Parallel and Perpendicular Lines (page 99)

1) $y = 2x$

2) $(y - 2) = -5(x + 1)$

3) $(y - 3) = 3(x - 2)$

a) $y - 1 = \frac{1}{2}x$

b) $y + 5 = -\frac{5}{3}(x + 2)$

c) $y - 3 = \frac{1}{3}(x - 1)$

Chapter 5 – Solving and Graphing Inequalities

Introduction to Inequalities (page 105)

1) Negative one is less than zero.

2) Three is greater than or equal to two.

3) Five is less than or equal to five.

4) x is greater than negative one and is less than or equal to three.

5) y is less than or equal to 4 and is greater than or equal to negative three.

6) 4 is greater than zero and is less than 10.

a) True

b) False

c) True

d) True

e) False

f) True

Solve an Inequality with Addition and Subtraction (page 107)

1) $x > 8$

2) $x \geq -7$

3) $x > 5$

4) $x \leq 2$

5) $x \leq -3$

6) $x < 6$

Solve an Inequality with Multiplication and Division (page 109)

1) $x > 5$

2) $x \geq -5$

3) $x \geq \frac{7}{4}$

4) $x > -30$

5) $x > 10$

6) $x \geq \frac{8}{9}$

Solve an Inequality in Several Steps (page 111)

1) $x < 9$

2) $x \leq -3$

3) $x \geq -\frac{1}{2}$

4) $x < 1$

5) $x > -2$

6) $x \geq -5$

Solve a Compound Inequality (page 113)

1) $-2 < x < 5$

2) $2 > x > 9$

3) $2 \geq x \geq -4$

4) $2 \geq x \geq 6$

5) $6 > x \geq -4$

6) $-4 \geq x > -1$

Solve an Inequality with Absolute Values (page 115)

1) $x < -2$ or $x > 2$

2) x can be any value.

3) x can be any value except zero.

4) $x \geq 4$ or $x \leq -4$

5) $x > 3$ or $x < -\dfrac{5}{3}$

6) $x \geq 6$ or $x \leq -2$

Solve an Inequality with Absolute Values (page 117)

1) $-5 < x < 5$

2) no solution

3) $x = 0$

4) $-3 \leq x \leq 3$

5) $-1 < x < 9$

6) $-\dfrac{2}{3} \leq x \leq 2$

Graph an Inequality with One Variable (page 119)

1)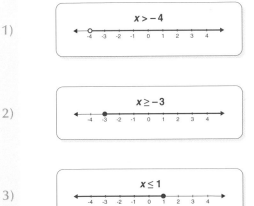

2)

3)

4)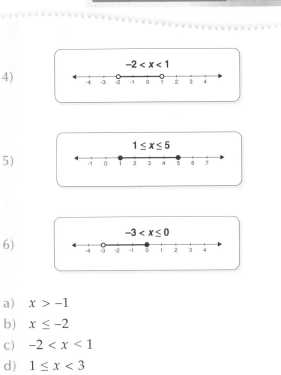

5)

6)

a) $x > -1$

b) $x \leq -2$

c) $-2 < x < 1$

d) $1 \leq x < 3$

Graph an Inequality with Two Variables (page 121)

1)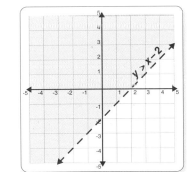

2)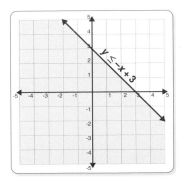

257

Practice Solutions

Chapter 5 — Solving and Graphing Inequalities *continued*

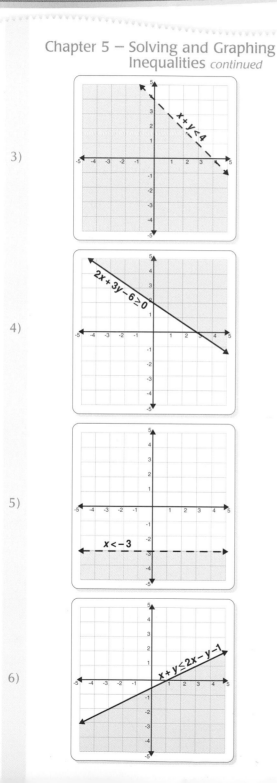

3)

4)

5)

6)

Chapter 6 — Working With Matrices

Introduction to Matrices (page 127)

Matrix A

Order = 3 x 2; element names: $a_{11} = 7$, $a_{12} = 5$, $a_{21} = -3$, $a_{22} = 8$, $a_{31} = 2$, $a_{32} = -6$

Matrix B

Order = 2 x 3; element names: $b_{11} = 5$, $b_{12} = 3$, $b_{13} = -2$, $b_{21} = -9$, $b_{22} = 6$, $b_{23} = 4$

Add and Subtract Matrices (page 129)

1)
$$C = \begin{bmatrix} 11 & -12 & -7 \\ -10 & -5 & 5 \\ 10 & -6 & -3 \end{bmatrix}$$

2)
$$D = \begin{bmatrix} 1 & 6 & -9 \\ -4 & 15 & -3 \\ -2 & 10 & -7 \end{bmatrix}$$

Multiply a Matrix by a Scalar (page 131)

1)
$$3A = \begin{bmatrix} -3 & 21 & -15 \\ 6 & 18 & 12 \\ 24 & -9 & -27 \end{bmatrix}$$

2)
$$-5A = \begin{bmatrix} 5 & -35 & 25 \\ -10 & -30 & -20 \\ -40 & 15 & 45 \end{bmatrix}$$

Multiply Matrices (page 135)

$$AB = \begin{bmatrix} 100 & 71 \\ 62 & 31 \\ 46 & 68 \end{bmatrix}$$

Chapter 7 — Solving Systems of Equations

Solve a System of Equations by Graphing (page 141)

1)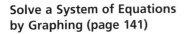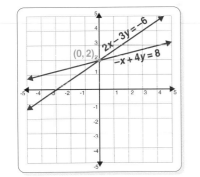

1)

2)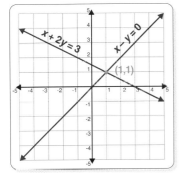

3)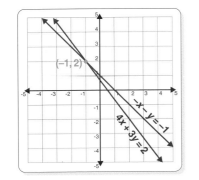

4)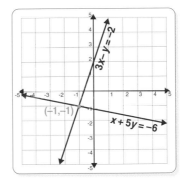

5)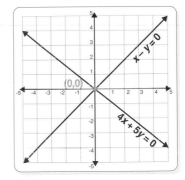

6)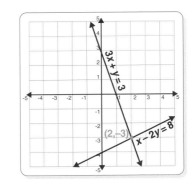

Practice Solutions

Chapter 7 — Solving Systems of Equations *continued*

Solve a System of Equations by Substitution (page 143)

1) $x = 1, y = 0$
2) $x = -1, y = 2$
3) $x = 2, y = 2$
4) $x = 0, y = 0$
5) $x = -2, y = -3$
6) $x = 0, y = -2$

Solve a System of Equations by Elimination (page 145)

1) $x = 3, y = -1$
2) $x = -17, y = 8$
3) $x = 3, y = -1$
4) $x = 1, y = 2$
5) $x = -1, y = -6$
6) $x = 2, y = -\dfrac{5}{3}$

Solve a System of Equations by Elimination (page 147)

1) $x = 1, y = -1$
2) $x = 0, y = 0$
3) $x = 2, y = 2$
4) $x = -1, y = 2$
5) $x = 3, y = 1$
6) $x = 1, y = 2$

Chapter 8 — Working With Polynomials

Introduction to Polynomials (page 153)

1) constant monomial
2) cubic binomial
3) linear polynomial with four terms
4) quadratic polynomial with five terms
5) quartic trinomial
6) quintic binomial

Multiplying Polynomials (page 155)

1) $x^2 + 6x + 8$
2) $3x^6 - 3x^5 + 6x^4$
3) $xy - x + y - 1$
4) $x^3 + x^2y - 5x^2$
5) $2x^2 + 5xy + 3y^2$
6) $x^2 - y^2$

Divide Polynomials Using Long Division (page 159)

1) $-x + 3 + \dfrac{x + 1}{x^2 + x + 1}$
2) $3x^2 - \dfrac{4x}{x^2 - 4}$
3) $2x^3 - \dfrac{2}{3x + 2}$
4) $x^3 + x$
5) $x^4 + 2x^3 - 2x^2 - 3x + 3 + \dfrac{10x^2 - 2}{x^3 + 2x + 1}$
6) $x - 1 + \dfrac{1}{x + 1}$

Divide Polynomials Using Synthetic Division (page 163)

1) $x^2 + 5x - 3 + \dfrac{1}{x - 2}$

2) $x^3 + 2x - 1 + \dfrac{1}{x + 1}$

3) $2x^3 - x + \dfrac{6}{x + 3}$

4) $-x^2 + 3x + 5$

5) $x^5 + x + 1 + \dfrac{3}{x - 1}$

6) $x^2 + 2 - \dfrac{10}{x + 5}$

Chapter 9 – Working With Radicals

Simplify Radicals (page 171)

1) 3

2) $2\sqrt[3]{10}$

3) $10\,|y|\,\sqrt{2y}$

4) $5x^2\sqrt{3x}$

5) Cannot be simplified.

6) $2\sqrt[4]{2}$

Add and Subtract Radicals (page 173)

1) $-4\sqrt{2}$

2) $11\sqrt[3]{10}$

3) $9\sqrt{3}$

4) $12\sqrt{3}$

5) $5\sqrt[3]{9}$

6) $4\sqrt[3]{2}$

Multiply Radicals (page 175)

1) $\sqrt{14}$

2) 11

3) $\sqrt[3]{24} = 2\sqrt[3]{3}$

4) $\sqrt{350} = 5\sqrt{14}$

5) $\sqrt[3]{20}$

6) $10\sqrt[4]{2}$

Divide Radicals (page 177)

1) $2\sqrt{2}$

2) $3\sqrt{4x} = 6\sqrt{x}$

3) $5\sqrt[3]{4x}$

4) $\dfrac{\sqrt{11}}{\sqrt{5}} = \dfrac{\sqrt{55}}{5}$

5) $\dfrac{3\sqrt{7}}{\sqrt{3}} = \dfrac{3\sqrt{21}}{3} = \sqrt{21}$

6) $\dfrac{7}{3}\sqrt{x^2 y^2} = \dfrac{7}{3}\,|x||y|$

Write Radicals as Exponents (page 179)

1) $2^{\frac{2}{3}}$

2) $(3x^2)^{\frac{1}{7}}$

3) $y^{\frac{2}{4}} - y^{\frac{1}{2}}$

4) $\sqrt[4]{5^3}$

5) $\sqrt[5]{x^2}$

6) $\sqrt[3]{y}$

Solve a Radical Equation (page 181)

1) $x = 26$

2) $x = 0$

3) $x = -30$

4) $x = 2$

5) $x = 517$

6) no solution

Practice Solutions

Chapter 10 — Factoring Expressions

Factor Using the Greatest Common Factor (page 187)

1) $2(1 + 2 + 3)$
2) $3(2 + 3 + 5)$
3) $6(2 + 3 + 10)$
4) $2(2y + x + 25z)$
5) $10(a + 10b + 5c)$
6) $7(a + b + 1)$

Factor Using the Greatest Common Factor (page 189)

1) $c(ab + ad + be)$
2) $x^2(1 + x + x^8)$
3) $x^2y^2(yz + x^2z^2 + xy^2)$
4) $a(1 + a^2 - a^4)$
5) $b(ac^3 - c + b^3)$
6) $-b^2(1 + b^2 - b^5)$ or $b^2(-1 - b^2 + b^5)$

Factor Using the Greatest Common Factor (page 191)

1) $2x(1 + 2x + 3x^2)$
2) $5x(2y + 5x + 7y^2)$
3) $4z^3(2 - 4z^2 + 5z^3)$
4) $9xyz(2z - 9xy^2)$
5) $-2(2a^2 + 5a^5 - 3a + 4)$
6) The greatest common factor is 1. The expression cannot be factored.

Factor by Grouping (page 193)

1) $(x + y)(a + b)$
2) $(x + y)(c - d)$
3) $(a + 2b)(3c - d)$
4) $(a + b)(2x + 3y)$
5) $(5x - y)(z + 1)$
6) $(x + 2)(y - 3)$

Recognizing Special Factor Patterns (page 196)

1) $(x \times 3)(x - 3)$
2) $(y^2 + a^2)(y^2 - a^2)$
   $= (y^2 + a^2)(y + a)(y - a)$
3) $(b^2 + 2)(b^2 - 2)$
4) $(x + y(x^2 - xy + y^2)$
5) $(a + 2)(a^2 - 2a + 4)$
6) $(b^2 + 3)(b^4 - 3b^2 + 9)$

Recognizing Special Factor Patterns (page 197)

1) $(x - y)(x^2 + xy + y^2)$
2) $(z - 1)(z^2 + z + 1)$
3) $(3 - 2a)(9 + 6a + 4a^2)$
4) $(2b - 1)(4b^2 + 2b + 1)$
5) $(x^2 - y^2)(x^4 + x^2y^2 + y^4)$
   $= (x \times y)(x - y)(x^4 + x^2y^2 + y^4)$
6) $(x^2 - 2)(x^4 + 2x^2 + 4)$

Factor Simple Trinomials (page 199)

1) $(x + 1)(x + 5)$
2) $(x + 2)(x + 10)$
3) $(x - 4)(x + 3)$
4) $(x - 2)(x - 3)$
5) $(x - 2)(x - 4)$
6) $(x + 9)(x - 1)$

Factor More Complex Trinomials (page 201)

1) $(2x - 1)(x + 3)$
2) $(x + 5)(x - 4)$
3) $(3x - 1)(2 - x)$
4) $(x + 1)(5x \quad 1)$
5) $(2x + 4)(3x - 3)$
6) $(3x + 1)(4x - 3)$

Chapter 11 – Solving Quadratic Equations and Inequalities

Solve Quadratic Equations by Factoring (page 207)

1) $x = 0$ or 7
2) $x = 0$ or -5
3) $x = 0$ or $\dfrac{8}{3}$
4) $x = 0$ or 10
5) $x = 0$ or $-\dfrac{1}{4}$
6) $x = 0$ or 4

Solve Quadratic Equations by Factoring (page 209)

1) $x = -1$ or 2
2) $x = -\dfrac{1}{2}$ or 3
3) $x = \dfrac{2}{3}$ or $\dfrac{5}{2}$
4) $x = 4$ or -5
5) $x = -\dfrac{1}{4}$ or 1
6) $x = 2$

Solve Quadratic Equations Using Square Roots (page 211)

1) $x = \pm 3$
2) $x = \pm 5$
3) $x = \pm 10$
4) $x = \pm 4$
5) $x = \pm 2$
6) $x = \pm 6$

Solve Quadratic Equations by Completing the Square (page 215)

1) $x - 1$ or -3
2) $x = 1$
3) no solution
4) $x = 1$ or -5
5) $x = 1$ or -1
6) $x = 2$ or -3

Practice Solutions

Chapter 11 – Solving Quadratic Equations and Inequalities *continued*

Solve Quadratic Equations Using the Quadratic Formula (page 217)

1) $x = 1$ or -2

2) $x = -3$ or 3

3) no solution

4) $x = \dfrac{1}{4}$ or $-\dfrac{2}{3}$

5) $x = 0$ or $\dfrac{7}{5}$

6) $x = 2$

Solve Cubic and Higher Equations (page 221)

1) $x = 1$, 2 or 3

2) $x = -1$ or 5

3) $x = 2$

4) $x = -\dfrac{1}{2}$ or 1

5) $x = 0$ or 1

6) $x = 0$, 1, -1 or -5

Solve Quadratic Inequalities (page 225)

1) $x < -1$ or $x > 3$

2) $x \neq 0$

3) $-5 < x < 1$

4) no solution

5) $1 \leq x \leq 2$

6) $-14 \leq x \leq -6$

Chapter 12 – Graphing Quadratic Equations

Introduction to Parabolas (page 231)

1) vertex $= (1,0)$, opens upwards

2) vertex $= \left(\dfrac{1}{3},-5\right)$, opens upwards

3) vertex $= (-1,2)$, opens downwards

4) vertex $= \left(0,\dfrac{3}{10}\right)$, opens upwards

5) vertex $= (0,2)$, opens downwards

6) vertex $= (2,-3)$, opens downwards

Write a Quadratic Equation in Vertex Form (page 233)

1) $y = (x - 1)^2 + 2$

2) $y = -(x - 2)^2 + 3$

3) $y = -2(x + 1)^2 - 1$

4) $y = (x - 3)^2 - 2$

5) This equation is already in vertex form. The two forms are the same if $b = 0$.

6) $y = (x + 5)^2 - 10$

Graph a Parabola (page 235)

1)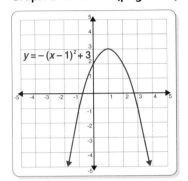

2)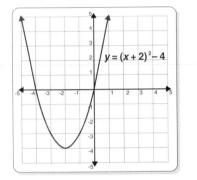

3)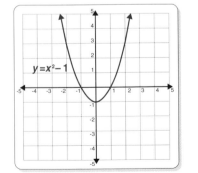

4)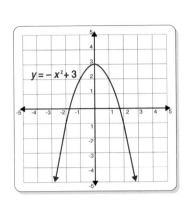

5)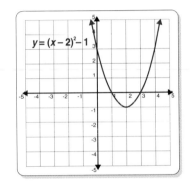

6)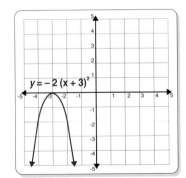

Test Your Skills Solutions

Chapter 1 — Algebra Basics

Question 1

a) 13
b) −4
c) −4
d) −6
e) −5
f) 1
g) 0
h) 5

Question 2

a) 2
b) 3
c) −4
d) 4
e) 14
f) −15
g) 20

Question 3

a) 3
b) 6
c) 0
d) −4
e) 9

Question 4

a) −13
b) 4
c) −10
d) −2
e) 1

Question 5

a) 4
b) $\dfrac{3}{2}$
c) $-\dfrac{1}{10}$
d) −1
e) $\dfrac{5}{4}$

Question 6

a) $3x^2 - 4x + 8$
b) $2x^4 + x^3 + 5x + 1$
c) $4x + 3xy - 5y$
d) $24x^3$
e) $7x$
f) $4x^2$

Question 7

a) x^2
b) $2x$
c) $-3x^6$
d) $3x - 2xy$
e) $-15 + 35x$
f) $-12x^3 + 8x^2 + 24xy$

Chapter 2 – Working With Fractions and Exponents

Question 1

a) 2

b) $\frac{2}{3}$

c) $-\frac{1}{5}$

d) $\frac{4}{5}$

e) $-\frac{2}{5}$

f) $\frac{2}{3}$

Question 2

a) $1\frac{1}{2}$

b) -2

c) $-6\frac{2}{3}$

d) $2\frac{1}{2}$

e) $1\frac{2}{5}$

Question 3

a) $\frac{7}{2}$

b) $\frac{7}{3}$

c) $-\frac{19}{9}$

d) $\frac{17}{4}$

e) $\frac{51}{10}$

Question 4

a) $\frac{1}{6}$

b) $\frac{6}{35}$

c) $-\frac{1}{36}$

d) $\frac{3}{4}$

e) $-\frac{7}{16}$

f) $\frac{1}{20}$

Question 5

a) $\frac{3}{6}$, $\frac{2}{6}$

b) $\frac{4}{20}$, $\frac{5}{20}$

c) $\frac{9}{12}$, $-\frac{10}{12}$

d) $-\frac{5}{50}$, $\frac{6}{50}$

e) $-\frac{13}{100}$, $\frac{40}{100}$

Question 6

a) $\frac{7}{10}$

b) $\frac{5}{4}$

c) $-\frac{3}{5}$

d) $\frac{5}{21}$

e) $\frac{9}{5}$

f) $-\frac{9}{8}$

Test Your Skills Solutions

Chapter 2 – Working With Fractions and Exponents *continued*

Question 7

a) 10^3

b) 10^{-3}

c) 2.39861×10^5

d) 1.023×10^4

e) 2.3×10^{-5}

f) 1.00013×10^6

g) 3×10^{-1}

Question 8

a) 5^6

b) 3^{-3}

c) $2^6 3^3$

d) 10^2

e) 6^{14}

f) $x^2 y^2$

g) $2^3 5^3$

Chapter 3 – Solving Basic Equations

Question 1

a) $x = 3$

b) $x = 14$

c) $x = -3$

d) $x = -4$

e) $x = \dfrac{5}{3}$

f) $x = 2$

g) $x = \dfrac{5}{8}$

h) $x = 24$

i) $x = 6$

j) $x = \dfrac{1}{2}$

k) $x = 7$

Question 2

a) $x = \dfrac{1}{3}$

b) $x = -1$

c) $x = -\dfrac{1}{2}$

d) $x = 2$

e) $x = -16$

f) $x = 1$

Question 3

a) $y = -x$

b) $y = 2$

c) $y = \dfrac{5}{2} - 2x$

d) $y = x - 5$

e) $y = \dfrac{x + 7}{5}$

f) $y = \dfrac{x + 3}{-5}$

Question 4

a) $x = -3, 3$

b) $x = -2, 2$

c) $x = 0$

d) no solution

e) $x = 1, 3$

f) $x = 7, 13$

g) $x = 0, -10$

Question 5

a) $x = -2, 2$

b) $x = -\dfrac{8}{3}, \dfrac{8}{3}$

c) no solution

d) $x = 2$

e) $x = 0, 1$

f) $x = \dfrac{4}{5}, 0$

g) $x = \dfrac{11}{3}, -3$

h) $x = \dfrac{1}{2}, -\dfrac{7}{2}$

Chapter 4 – Graphing Linear Equations

Question 1

a)

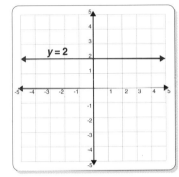

b)

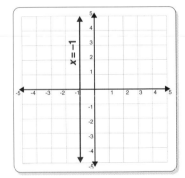

c)

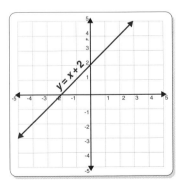

269

Chapter 4 – Graphing Linear Equations *continued*

d)

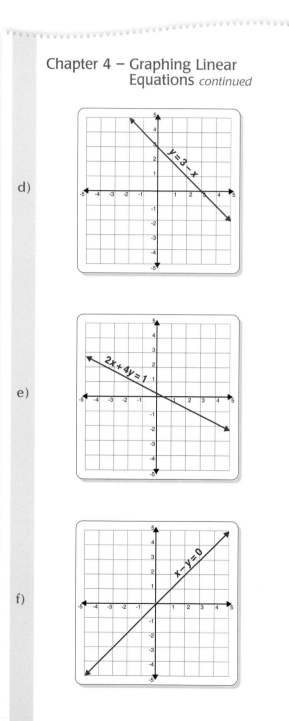

e)

f)

g)

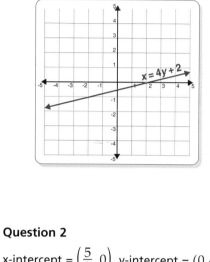

Question 2

a) x-intercept = $\left(\frac{5}{2}, 0\right)$, y-intercept = $(0, -5)$

b) x-intercept = $(2, 0)$, y-intercept = $(0, 6)$

c) x-intercept = $(1, 0)$, y-intercept = $(0, -1)$

d) x-intercept = $(0, 0)$, y-intercept = $(0, 0)$

e) x-intercept = $\left(-\frac{1}{2}, 0\right)$, y-intercept = $\left(0, -\frac{1}{7}\right)$

f) x-intercept = $(-4, 0)$, y-intercept = $(0, 6)$

Question 3

a) slope = 3

b) slope = -4

c) slope = 0

d) slope is undefined.

e) slope = 1

f) slope = $-\frac{2}{5}$

Chapter 5 – Solving and Graphing Inequalities

Question 1

a) $x > 3$

b) $x < 4$

c) $y \leq -5$

d) $x \geq 12$

e) $x < 0$

f) $x < -5$

g) $y \leq -3$

h) $x < 3$

i) $x < 0$

Question 4

a) $y = -x - 2$, slope $= -1$, y-intercept $= (0, -2)$

b) $y = \dfrac{3x}{2} + \dfrac{5}{2}$, slope $= \dfrac{3}{2}$, y-intercept $= \left(0, \dfrac{5}{2}\right)$

c) $y = \dfrac{3x}{4} - 6$, slope $= \dfrac{3}{4}$, y-intercept $= (0, -6)$

d) $y = x - 1$, slope $= 1$, y-intercept $= (0, -1)$

e) $y = 5x - 2$, slope $= 5$, y-intercept $= (0, -2)$

f) $y = \dfrac{2x}{3} - \dfrac{8}{3}$, slope $= \dfrac{2}{3}$, y-intercept $= \left(0, -\dfrac{8}{3}\right)$

Question 2

a) $-1 < x \leq 3$

b) $7 < x < 11$

c) $-4 \leq x \leq -1$

d) $\dfrac{2}{3} < x < \dfrac{4}{3}$

e) $\dfrac{3}{2} < x \leq 3$

f) $\dfrac{5}{2} \leq x \leq \dfrac{7}{2}$

g) $-5 < x < -1$

Question 5

a) $y = x + 2$

b) $y = -2x + 25$

c) $y = 4x + 12$

d) $y = 6x$

e) $y = -\dfrac{1x}{2} - 1$

f) $y = \dfrac{5x}{4} - 5$

Question 3

a) $-2 < x < 2$

b) $x < -3$ or $x > 3$

c) $-5 < x < 5$

d) $-3 \leq x \leq 5$

e) $x \leq -2$ or $x \geq -1$

f) $\dfrac{2}{5} < x < \dfrac{18}{5}$

g) no solution

Question 6

a) parallel

b) perpendicular

c) perpendicular

d) neither

e) parallel

f) perpendicular

Chapter 5 – Solving and Graphing Inequalities *continued*

Question 4

a)
$x > 3$

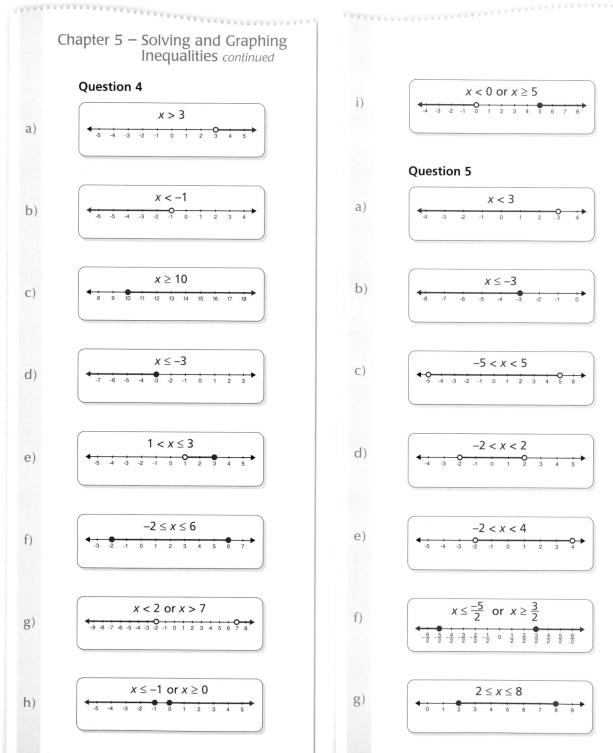

b)
$x < -1$

c)
$x \geq 10$

d)
$x \leq -3$

e)
$1 < x \leq 3$

f)
$-2 \leq x \leq 6$

g)
$x < 2$ or $x > 7$

h)
$x \leq -1$ or $x \geq 0$

i)
$x < 0$ or $x \geq 5$

Question 5

a)
$x < 3$

b)
$x \leq -3$

c)
$-5 < x < 5$

d)
$-2 < x < 2$

e)
$-2 < x < 4$

f)
$x \leq \dfrac{-5}{2}$ or $x \geq \dfrac{3}{2}$

g)
$2 \leq x \leq 8$

Question 6

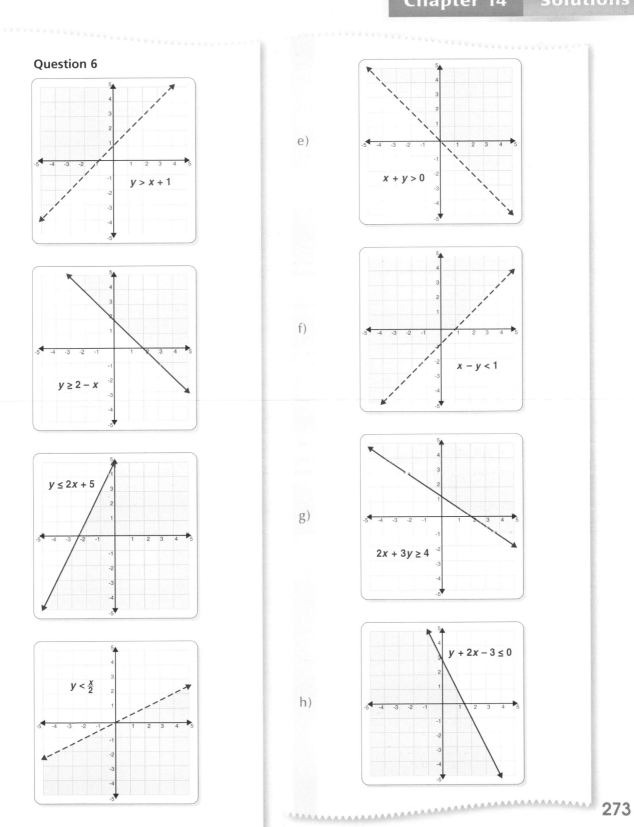

a) $y > x + 1$

b) $y \geq 2 - x$

c) $y \leq 2x + 5$

d) $y < \dfrac{x}{2}$

e) $x + y > 0$

f) $x - y < 1$

g) $2x + 3y \geq 4$

h) $y + 2x - 3 \leq 0$

Test Your Skills Solutions

Chapter 6 — Working With Matrices

Question 1

A = 2 x 3

B = 3 x 2

C = 2 x 2

Question 2

$$A + B = \begin{bmatrix} -1 & 14 \\ -3 & 7 \end{bmatrix}$$

A + C = undefined

$$C + D = \begin{bmatrix} 3 & 0 & 7 \\ 6 & -12 & 19 \end{bmatrix}$$

$$B - A = \begin{bmatrix} -5 & 6 \\ 3 & -3 \end{bmatrix}$$

D − B = undefined

Question 3

a)   AD, defined, order = 2 x 3

b)   AB, undefined

c)   CD, defined, order = 3 x 3

d)   DC, defined, order = 2 x 2

e)   DA, undefined

f)   DB, defined, order = 2 x 3

Question 4

$$2A = \begin{bmatrix} 4 & 6 \\ -2 & 8 \\ -4 & 10 \end{bmatrix}$$

$$-3B = \begin{bmatrix} 0 & 12 \\ -3 & -3 \\ -6 & 9 \end{bmatrix}$$

$$10C = \begin{bmatrix} 30 & 100 \\ -20 & 50 \end{bmatrix}$$

$$2B = \begin{bmatrix} 0 & -8 \\ 2 & 2 \\ 4 & -6 \end{bmatrix}$$

Question 5

$$AB = \begin{bmatrix} 0 & 0 \\ 0 & 0 \end{bmatrix}$$

$$BA = \begin{bmatrix} 0 & -11 \\ 0 & 0 \end{bmatrix}$$

$$AC = \begin{bmatrix} 0 & 4 & 10 \\ 0 & -6 & -15 \end{bmatrix}$$

BD = undefined

$$CD = \begin{bmatrix} 15 & 0 & 14 \\ -7 & 9 & -10 \end{bmatrix}$$

DC = undefined

Chapter 7 — Solving Systems of Equations

Question 1

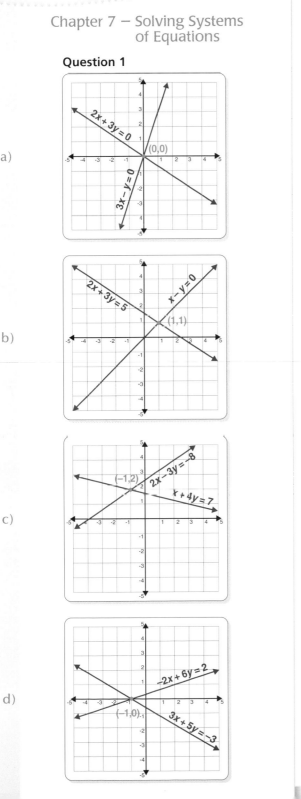

a)

b)

c)

d)

e)

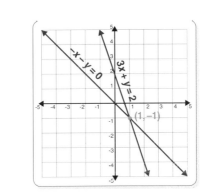

Question 2

a) $x = 2, y = 5$

b) $x = 0, y = 0$

c) $x = 3, y = -1$

d) $x = -2, y = -2$

e) $x = 1, y = 0$

Question 3

a) $x = 1, y = 1$

b) $x = 2, y = 3$

c) $x = 0, y = -1$

d) $x = 2, y = -1$

e) $x = 3, y = 3$

Question 4

a) $x = 1, y = -1$

b) $x = 3, y = 4$

c) $x = -1, y = -1$

d) $x = 1, y = 0$

e) $x = -2, y = -2$

Test Your Skills Solutions

Chapter 8 – Working With Polynomials

Question 1

a) degree = 3

b) degree = 4

c) degree = 6

d) degree = 7

e) degree = 1

f) degree = 2

Question 2

a) $x^2 + 5x + 6$

b) $2x^2 - 3x - 2$

c) $x^2 - 25$

d) $2x^3 - 9x^2 + x + 12$

e) $x^3 + 2x^2 - 8x$

f) $2x^4 + 5x^3 - 2x^2 - 4x - 1$

Question 3

a) $x^2 + 3x + 4$

b) $3x^3 - 2x^2 + 1 - \dfrac{1}{x + 1}$

c) $-x^2 + x + 5 + \dfrac{3}{2x - 1}$

d) $2x^2 - 8 + \dfrac{2}{x^2 + 1}$

e) $x^2 + x + 1 + \dfrac{x + 1}{x^2 + x - 2}$

Question 4

a) $x^2 + x + 1$

b) $2x^2 + x - 3 + \dfrac{6}{x + 2}$

c) $x^3 - x^2 + 2x + 5 - \dfrac{5}{x + 5}$

d) $3x^3 - 2x^2 - x + 7 + \dfrac{10}{x - 3}$

e) $5x^4 - x + 10$

Chapter 9 – Working With Radicals

Question 1

a) 2

b) $5\sqrt{2}$

c) 3

d) $2\sqrt[4]{2}$

e) $4\sqrt{3}$

f) $|x|$

g) y

h) $|z|\sqrt{z}$

Question 2

a) $6\sqrt{2}$

b) $5\sqrt[3]{5}$

c) $2\sqrt{2} + 10\sqrt{3}$

d) $3\sqrt{3}$

e) $-3\sqrt[3]{5} + 10\sqrt{5}$

Question 3

a) $\sqrt{35}$

b) 2

c) $2\sqrt[3]{2}$

d) $\sqrt[5]{80}$

e) x

Question 4

a) $\sqrt{2}$

b) $\sqrt{3}$

c) $\sqrt[3]{20}$

d) $\sqrt[5]{7}$

e) x^2

Question 5

a) $5^{\frac{1}{2}}$

b) $10^{\frac{1}{3}}$

c) $3^{\frac{2}{5}}$

d) $x^{\frac{1}{2}}$

e) $y^{\frac{z}{4}}$

Question 6

a) $\sqrt{6}$

b) $\sqrt{7}$

c) $\sqrt[4]{z}$

d) $\sqrt[5]{x^4}$

e) $3\sqrt[4]{3}$

Question 7

a) $x = 9$

b) no solution

c) $x = 4$

d) $x = \dfrac{125}{4}$

e) $x = 1$

Test Your Skills Solutions

Chapter 10 – Factoring Expressions

Question 1

a) $3(1 + 2 + 5)$

b) $4(3 + 2 + 5)$

c) $2x(4x + 1 + 3x^2)$

d) $abc(a + b + c)$

e) $3a^2b(2ab + 3 + 7a^2b^2)$

Question 2

a) $(x + 4)(x + 3)$

b) $(2x + 3)(2x - 1)$

c) $(3y + 5)(y - 4)$

d) $(x + 1)(x + 2)$

e) $(y + 1)(x - 3)$

f) $(x + y)(2y - x)$

Question 3

a) $(x + 1)(x + 4)$

b) $(x - 2)(x + 3)$

c) $(z - 5)(z + 2)$

d) $(y - 3)(y + 3)$

e) $(a + 2)(a + 7)$

Question 4

a) $(2x - 1)(x + 2)$

b) $(3x + 2)(2x + 1)$

c) $(x - 4)(3x + 5)$

d) $(2x - 5)(2x - 1)$

e) $(5x - 1)(4x + 3)$

Question 5

a) $(x - y)(x + y)$

b) $(a - b)(a^2 + ab + b^2)$

c) $(x + 2)(x^2 - 2x + 4)$

d) $(y - 1)(y^2 + y + 1)$

e) $(x^2)^2 - 1^2 = (x^2 + 1)(x^2 - 1)$
$$= (x^2 + 1)(x + 1)(x - 1)$$

Chapter 11 – Solving Quadratic Equations and Inequalities

Question 1

a) $x = 1$ or -4

b) $x = 0$ or -3

c) $x = \dfrac{1}{2}$ or 3

d) $x = -\dfrac{3}{4}$ or $-\dfrac{1}{2}$

e) $x = 2$

Question 2

a) $x = 4$ or -4

b) $x = 2$ or -2

c) $x = 5$ or -5

d) $x = \sqrt{6}$ or $-\sqrt{6}$

e) $x = 0$

Question 3

a) $x = 0$ or 4

b) $x = -1$

c) $x = 0$ or -6

d) $x = -2$ or 4

e) $x = \dfrac{3}{2}$ or $-\dfrac{1}{2}$

Question 4

a) $x = 1$ or 2

b) $x = 0$ or -3

c) $x = -\dfrac{1}{2}$ or -1

d) $x = -\dfrac{2}{3}$ or $\dfrac{1}{4}$

e) $x = \dfrac{(-5 + \sqrt{17})}{4}$ or $\dfrac{(-5 - \sqrt{17})}{4}$

Question 5

a) $x = 1$

b) no solution

c) $x = 1, -1$

d) $x = 1, 3, \dfrac{2}{3}$

e) $x = 1, -1, 3, -3$

Question 6

a) $x < -2$ or $x > 1$

b) all x not equal to 0

c) $x = 0$

d) $-3 < x < 2$

e) $x \leq 0$ or $x \geq 10$

f) $\dfrac{1}{2} \leq x \leq \dfrac{2}{3}$

Chapter 12 – Graphing Quadratic Equations

Question 1

a)   vertex = $(1,2)$, opens upward

b)   vertex = $(-6,-3)$, opens downward

c)   vertex = $\left(-4, \dfrac{1}{2}\right)$, opens upward

d)   vertex = $\left(\dfrac{1}{3}, 0\right)$, opens downward

e)   vertex = $(50,-75)$, opens upward

Question 2

a) $y = (x - 1)^2$, vertex = $(1,0)$,
     opens upward

b) $y = (x + 3)^2 - 4$, vertex = $(-3,-4)$,
     opens upward

c) $y = -(x + 1)^2 + 3$, vertex = $(-1,3)$,
     opens downward

d) $y = -4\left(x - \dfrac{1}{2}\right)^2 + 4$, vertex = $\left(\dfrac{1}{2}, 4\right)$,
     opens downward

e) $y = 25\left(x + \dfrac{2}{5}\right)^2 + 12$, vertex = $\left(-\dfrac{2}{5}, 12\right)$,
     opens upward

Question 3

a)

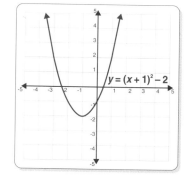

b)

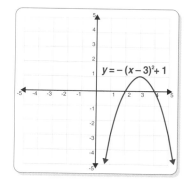

c)

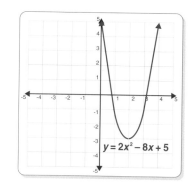

d)

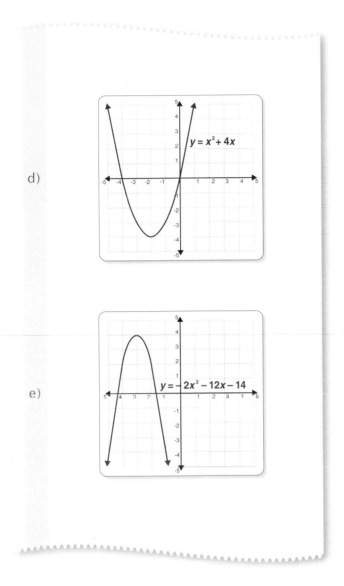

e)

Index

Index

Index

Index

Index

GUITAR

MARAN ILLUSTRATED™ Guitar is an excellent resource for people who want to learn to play the guitar, as well as for current musicians who want to fine tune their technique. This full-color guide includes over 500 photographs, accompanied by step-by-step instructions that teach you the basics of playing the guitar and reading music, as well as advanced guitar techniques. You will also learn what to look for when purchasing a guitar or accessories, how to maintain and repair your guitar and much more.

Whether you want to learn to strum your favorite tunes or play professionally, MARAN ILLUSTRATED™ Guitar provides all the information you need to become a proficient guitarist.

BOOK BONUS!

Visit **www.maran.com/guitar** to download MP3 files you can listen to and play along with for all the chords, scales, exercises and practice pieces in the book.

ISBN: 1-59200-860-7
Price: $24.99 US; $33.95 CDN
Page count: 320

PIANO

MARAN ILLUSTRATED™ Piano is an information-packed resource for people who want to learn to play the piano, as well as current musicians looking to hone their skills. Combining full-color photographs and easy-to-follow instructions, this guide covers everything from the basics of piano playing to more advanced techniques. Not only does MARAN ILLUSTRATED™ Piano show you how to read music, play scales and chords and improvise while playing with other musicians, it also provides you with helpful information for purchasing and caring for your piano. You will also learn what to look for when you buy a piano or piano accessories, how to find the best location for your piano and how to clean your piano.

ISBN: 1-59200-864-X

Price: $24.99 US; $33.95 CDN

Page count: 304

DOG TRAINING

MARAN ILLUSTRATED™ Dog Training is an excellent guide for both current dog owners and people considering making a dog part of their family. Using clear, step-by-step instructions accompanied by over 400 full-color photographs, MARAN ILLUSTRATED™ Dog Training is perfect for any visual learner who prefers seeing what to do rather than reading lengthy explanations.

Beginning with insights into popular dog breeds and puppy development, this book emphasizes positive training methods to guide you through socializing, housetraining and teaching your dog many commands. You will also learn how to work with problem behaviors, such as destructive chewing, excessive barking and separation anxiety.

ISBN: 1-59200-858-5

Price: $19.99 US; $26.95 CDN

Page count: 256

KNITTING & CROCHETING

MARAN ILLUSTRATED™ Knitting & Crocheting contains a wealth of information about these two increasingly popular crafts. Whether you are just starting out or you are an experienced knitter or crocheter interested in picking up new tips and techniques, this information-packed resource will take you from the basics, such as how to hold the knitting needles or crochet hook and create different types of stitches, to more advanced skills, such as how to add decorative touches to your projects and fix mistakes. The easy-to-follow information is communicated through clear, step-by-step instructions and accompanied by over 600 full-color photographs—perfect for any visual learner.

This book also includes numerous easy-to-follow patterns for all kinds of items, from simple crocheted scarves to cozy knitted baby outfits.

ISBN: 1-59200-862-3

Price: $24.99 US; $33.95 CDN

Page count: 304

maran illustrated

WEIGHT TRAINING

MARAN ILLUSTRATED™ Weight Training is an information-packed guide that covers all the basics of weight training, as well as more advanced techniques and exercises.

MARAN ILLUSTRATED™ Weight Training contains more than 500 full-color photographs of exercises for every major muscle group, along with clear, step-by-step instructions for performing the exercises. Useful tips provide additional information and advice to help enhance your weight training experience.

MARAN ILLUSTRATED™ Weight Training provides all the information you need to start weight training or to refresh your technique if you have been weight training for some time.

ISBN: 1-59200-866-6
Price: $24.99 US; $33.95 CDN
Page count: 320

YOGA

MARAN ILLUSTRATED™ Yoga provides a wealth of simplified, easy-to-follow information about the increasingly popular practice of Yoga. This easy-to-use guide is a must for visual learners who prefer to see and do without having to read lengthy explanations.

Using clear, step-by-step instructions accompanied by over 500 full-color photographs, this book includes all the information you need to get started with yoga or to enhance your technique if you have already made yoga a part of your life. MARAN ILLUSTRATED™ Yoga shows you how to safely and effectively perform a variety of yoga poses at various skill levels, how to breathe more efficiently, how to customize your yoga practice to meet your needs and much more.

ISBN: 1-59200-868-2

Price: $24.99 US; $33.95 CDN

Page count: 320

Did you like this book? MARAN ILLUSTRATED™ offers books on the most popular computer topics, using the same easy-to-use format of this book. We always say that if you like one of our books, you'll love the rest of our books too!

Here's a list of some of our best-selling computer titles:

Guided Tour Series - 240 pages, Full Color

MARAN ILLUSTRATED's Guided Tour series features a friendly disk character that walks you through each task step by step. The full-color screen shots are larger than in any of our other series and are accompanied by clear, concise instructions.

	ISBN	Price
MARAN ILLUSTRATED™ Computers Guided Tour	1-59200-880-1	$24.99 US/$33.95 CDN
MARAN ILLUSTRATED™ Windows XP Guided Tour	1-59200-886-0	$24.99 US/$33.95 CDN

MARAN ILLUSTRATED™ Series - 320 pages, Full Color

This series covers 30% more content than our Guided Tour series. Learn new software fast using our step-by-step approach and easy-to-understand text. Learning programs has never been this easy!

	ISBN	Price
MARAN ILLUSTRATED™ Access 2003	1-59200-872-0	$24.99 US/$33.95 CDN
MARAN ILLUSTRATED™ Computers	1-59200-870-4	$24.99 US/$33.95 CDN
MARAN ILLUSTRATED™ Excel 2003	1-59200-876-3	$24.99 US/$33.95 CDN
MARAN ILLUSTRATED™ Mac OS® X v.10.4 Tiger™	1-59200-878-X	$24.99 US/$33.95 CDN
MARAN ILLUSTRATED™ Office 2003	1-59200-890-9	$29.99 US/$39.95 CDN
MARAN ILLUSTRATED™ Windows XP	1-59200-870-4	$24.99 US/$33.95 CDN

101 Hot Tips Series - 240 pages, Full Color

Progress beyond the basics with MARAN ILLUSTRATED's 101 Hot Tips series. This series features 101 of the coolest shortcuts, tricks and tips that will help you work faster and easier.

	ISBN	Price
MARAN ILLUSTRATED™ Windows XP 101 Hot Tips	1-59200-882-8	$19.99 US/$26.95 CDN